The Inconvenient Skeptic

The Comprehensive Guide to the Earth's Climate

John Kehr

The Inconvenient Skeptic: The Comprehensive Guide to the Earth's Climate
Copyright © 2011 by John H. Kehr

ISBN 978-0-9847829-1-8 (paperback)

Dedicated to Jenny:

On so many levels this book would not
exist without you or your support.

Table of Contents

Chapter 1   Introduction

I am a global warming skeptic. I am in fact very skeptical about the whole idea that carbon dioxide (CO_2) can cause global warming. The funny part is that I was not always a skeptic. A few short years ago I wasn't a skeptic at all. It was my research into the science of global warming that turned me into a skeptic. The conclusion I reached from my research was that atmospheric CO_2 plays a very minor role in regulating the Earth's temperature.

I do need to clarify what I mean by global warming as well. The Earth has experienced several periods of warming in the past 200 years. There have also been periods when the Earth was cooling and other periods where the temperature was changing very little. Overall there have been more periods of warming. The result is that the Earth today is warmer than it was 200 years ago. I fully agree that this has happened and the Earth is warmer now than it was in the 1800's.

I am skeptical that CO_2 played a significant role in causing that change in temperature. So when I say I am skeptical of global warming, I am very specifically discussing the theory that increasing CO_2 levels will cause the Earth's temperature to increase. That is the foundation of **AGW (Anthropogenic Global Warming)**. This theory states that humanity is causing the warming that the Earth has recently experienced by increasing the concentration of CO_2 in the atmosphere through the burning of fossil fuels.

I never intended to write a book about global warming. If you had asked me 4 years ago about global warming I would have answered that CO_2 levels would probably make a difference, but the dangers were overstated. In my discussion with many other engineers that is a fairly normal position. It was certainly the view I had before I began my research.

Since I thought the doom and gloom predictions were overstated I simply didn't care about the issue. I knew that many people were getting worked up about it, but it didn't reach a high level of importance for me. I saw no harm in wanting more fuel efficient cars or more efficient appliances in a person's home. For me it fell into the "no harm, no foul" category. There was no clear danger from the solutions so even if the risks were overstated, the solutions were not a risk either. That is a good recipe for being apathetic about the issue as a whole and that was how I felt about it.

As a result I didn't take the time to research the science behind global warming. There was very little reason for me to spend my time studying something that didn't seem to matter; so I didn't. I do like to read science articles and occasionally I would come across an interesting article that seemed to support global warming or weaken the case against it. That did nothing to make me want to learn more about it.

That was a serious mistake on my part and it continues to be a serious mistake by many in the scientific community. Some of the information being presented as science is absurd. One example of this is a video that portrays CO_2 as a bunch of thugs mugging sunbeams trying to escape the Earth[1]. While some will argue that such analogies are helpful in trying to simplify the issue, the real effect is that such misrepresentations prevent people from understanding the issue. As a science junky I am offended by attempts to mislead people.

I have always been a science junky. I was one of those kids that wanted to be an astronaut when I grew up. At the age of 8 I got up at 3am to watch the very first space shuttle launch from t-minus 2 hours and 20 minutes. There is a very embarrassing video of me when I was about 10 years old talking about how I was going to be a scientist/astronaut when I grew up. It never even crossed my mind to do something different.

So it should be no surprise that I got my degree in Chemical Engineering from Kansas State University in Manhattan, Kansas. I will disclose at this point that I did work one summer as an intern working in the natural gas fields in Texas. I disliked that job and after working there I was even more determined to work in the field I really wanted to be in which was semiconductors. Shortly before graduating I got the job I wanted which was working in an R&D semiconductor Fab. It is where I have been for more than 12 years now.

In that time I have been involved in more than 10 generations of semiconductor technology. I have seen first hand just how far science can take technology. Being an engineer for the private sector requires results. The technology that has been developed in the past 10 years is breathtaking because of the research and development that has taken place. Things that were not even dreams 10 years ago are commonplace today. The precision of the technology in the semiconductor field today is stunning. That is exactly why I became an engineer and why I have continued in that field.

Those are my credentials. Research is something I have lots of exposure to. I have seen new ideas explored, tested and implemented into the latest technology with regular frequency. I have also seen new ideas discarded for not being the technology that was needed. Good ideas come and go, but the technology that I have been involved in must always be smaller, faster and cheaper. There is no time to waste on bad ideas, so I am used to quickly discarding bad ideas. In my view it is time to discard the theory of Anthropogenic Global Warming (AGW).

So how did a Research and Development Engineer at a major semiconductor company get involved enough in the issue of global warming to write a book about it? Especially when only a few years ago I was apathetic about the entire subject? That is a good question and it is an important question because the motivation of everyone in the global warming debate has become suspect.

This is because the global warming debate has become extremely political. Accusations fly

---

1   As shown in the movie *An Inconvenient Truth*.

regularly that anyone who is a skeptic is just a tool of big oil companies. The fact that I interned for an oilfield service company back in the summer of 1998 will be used as proof that I am still working for big oil. I find that amusing since I rejected the ensuing job offer because I didn't want to work for the oil industry. Since that was the first professional job offer I had ever received, it was scary to reject it, but I was determined to find a job in the semiconductor industry. Interning for the oil industry had only strengthened my resolve to not work in that field.

That is why I am going to tell the story of why I decided to research the topic of global warming. I did not start off with any agenda or any goal of proving it wrong or right. I simply reached a point where I decided that I wanted to understand the issue. Everything that I have learned and discovered along the way is simply the result of my desire to understand what drives the Earth's climate.

It all started when I met my wife, Jenny. She is a wonderful person who at the time was committed to living the green life and doing everything she could to reduce CO_2 emissions. This did not bother me at all, but neither was it something that I was very concerned about. I remained apathetic about it despite her best efforts.

I am a strong supporter of fuel efficiency for many reasons that have nothing to do with CO_2 emissions. I am opposed to pollution and my Eagle Scout project was cleaning up an illegal dumping site. Some construction crew had dumped all the residual concrete and other materials in an empty field a ways away from an apartment complex that they built. I cleaned it up. Taking action is one of those things I seem to do when I find something that I can make better.

As an engineer I understand how dangerous pollution can be and absolutely support being responsible in preventing pollution and cleaning it up when accidents do happen. Being skeptical about global warming does not mean I do not care about the Earth. I care a great deal about the Earth and I believe that humanity is learning to do a better job at reducing pollution. There is further to go, but great progress has been made in the past 30 years.

Despite the fact that I care about pollution and favor fuel efficiency, I am now an ardent global warming skeptic. How can that be? The answer is simple. I decided to look at the science behind global warming. It didn't happen overnight and it took almost a year before I was actively researching the topic, but it all started with simple request from Jenny for me to watch Al Gore's movie *An Inconvenient Truth*. That is not the type of movie I would normally watch, but she wanted me to watch it.

I will admit that I was more than a little skeptical of watching a scientific movie about global warming that was made by Al Gore. Al Gore and science go together like Spongebob Squarepants and Mensa. In the end, I decided it would be a good thing for me to know more about global warming than I did at the time. I told Jenny I would watch the movie after I did some of my own research into the topic.

That was a truly profound decision, even if I didn't know it at the time. Getting the facts about the science of global warming is comparable to untying the Gordian Knot. Each side of the debate is very well entrenched and they don't even discuss science anymore. It has devolved into a series of talking points that each side memorizes for every possible comment that could be made. There is no real thought process behind most discussion anymore, just mindless regurgitation.

Somewhere in the mix of it all there is good information, but it has become buried deep inside misinformation and is scattered about and disorganized. There is simply no place where someone can go to get a good scientific overview of the Earth's climate. That is all I wanted to find, but everything that was available was just arranged as talking points designed to highlight the flaws in the other side's arguments. That is not science, that is political debate.

I let Jenny know that it was going to take a little longer than expected because I had made my second decision on the issue. I was going to have to analyze the raw data and analyze it like I would something at work. I had decided to look at the whole issue and reach an independent conclusion about global warming. That was my equivalent solution to cutting the Gordian Knot. Instead of sorting through the conclusions of other people, I would look at the data only and reach my own conclusion.

That is something that I am not sure has happened before. I considered the theory of global warming possible, but I was not convinced that the consequences were being accurately presented to people. That was my view when I started to earnestly research the topic. I did not start out disbelieving the whole thing and trying to prove it wrong, I simply wanted to understand the issue fully so I could determine how seriously to take the potential risks. That was what I expected to learn from my research.

It didn't cross my mind initially that the entire theory itself was fundamentally incorrect, but that is exactly the conclusion that I eventually reached. Not only that, but the actions that are being proposed to counter global warming are becoming increasingly dangerous. The dangers range from economy-wrecking laws which will radically increase the price of energy to putting mirrors in space to reflect sunlight away from the Earth. I cannot even express how amazingly stupid that last idea is.

The situation had been changing all along, but because I had been apathetic about it, I was not aware of the increasingly absurd solutions that were being considered. The situation is far different from what it had first appeared to be once I really started to look at the issues. It isn't simply a case of no harm, no foul like I once thought it was.

Understanding the Earth's climate and the potential of global warming has stopped being a scientific curiosity to me and has become a serious topic because the solutions being proposed would be detrimental to my family's life. If the threat of global warming was as real as projected, then it would undoubtedly be worth the sacrifices that would be required. My children are important to me and I would never sacrifice their future.

4

Let me be very clear on this point. If I had reached the conclusion that global warming was a threat, then I would have dedicated my engineering experience to solutions to reduce CO_2 emissions. That would be simple for me to accomplish as solar power is related to my field of experience and it would be a quick transition for me to make. Nor would it be difficult for me to get involved with nuclear power which is a much better solution for generating electricity.

I would also be trying hard to convince people that global warming was real and should be taken seriously. It was the scientific conclusion that I reached that decided the course of action I have now taken. Because I took the time to understand the issue from the ground up, I have a different perspective than most people involved in the debate. I have an engineer's perspective on global warming.

Once I had taken the time to understand the issue I was left with another problem. I had a solid understanding of the issue from a very practical point of view. I could see how the information I had could benefit many other people. That left me with the responsibility to share what I had discovered. I was able to put together a comprehensive picture of the Earth's climate and how CO_2 fit into the picture.

That is what this book contains. It isn't even really about global warming. It is about understanding the Earth's climate. It does discuss global warming because there is a part of the Earth's climate that CO_2 does play an important role. Understanding the climate as a whole is much more important than simply a discussion about CO_2. Once the climate as a whole is understood, it is easy to reach a conclusion about what the increasing concentration of CO_2 will have on the Earth's climate.

It is possible from what I present that some people will reach a different conclusion from the one I have reached. It does happen that scientists can reach different conclusions from the same data. One current day example of this are the two very different conclusions reached about the velociraptor dinosaur that was made famous by the movie *Jurassic Park*.

The debate is about what type of hunters the velociraptors were. The movie supports the theory that they were pack hunters. It shows them working together much like a pack of wolves work together today. The other theory is that they were solitary scavengers that fought each other when scavenging large creatures, behavior more like modern day komodo dragons. These are two very different conclusions reached from the exact same data by the paleontologists who study the velociraptors.

I expect that people will reach different conclusions from my own. That is how science is supposed to work in the real world. Regardless of what a person believes about global warming, this book does provide an organized overview of the Earth's climate and the different factors that drive it. I hope that even those who reach a different conclusion will agree that it does provide a good overview of the science and the debate that surrounds global warming.

There are a few things that I hope each reader will gain from reading this book. First is a better understanding of the factors that drive the climate. While the Earth is large and the climate is complicated, understanding the most important aspects is fairly easy to understand. These factors have driven the climate for millions of years and they are not going to change.

The second item and perhaps the most important thing that I want readers to gain is an understanding that the Earth is always changing. Not once in the past 50 million years has the Earth's climate stayed the same for any significant amount of time. The third is directly tied to the second one. It is to understand that the Earth will change in the future, no matter what mankind does.

Regardless of what conclusion you reach about global warming, understanding that the Earth will change in the future will be a new perspective for some people. There is no chance that the Earth as it is today, will stay as it is today. The ocean levels will be different, the temperature will be different and how we interact with the Earth will be different. The Earth does not stay the same. The only thing that mankind controls is how prepared we are for the changes that will happen.

The real debate about global warming isn't even about the Earth changing because the Earth is always changing. The real debate is about how much influence will the level of CO_2 in the atmosphere have on the natural changes that will take place in the Earth's climate regardless of what mankind does.

People already accept that January is always different from July. That is true regardless of where you live in the world. The new perspective is that the July that we experience today is different than the July that happened 9,000 years ago (warmer than today) and that was different from July 20,000 years ago (much, much colder than today). It should be no surprise that in 5,000 years July will be different from what it is today.

Simply accepting that such changes have always happened and are natural requires a different view of the world. If a person expects things to stay the same, then they will always be surprised when change happens. If one understands that the Earth will change, then they are prepared when the inevitable changes do take place.

It is being able to predict how the Earth will change in the future that is really important about the climate debate. There are only two options. If the theory of global warming is correct, then the Earth will warm up in the manner predicted by warmists (those that believe in global warming). The other option is that the Earth will not be significantly influenced by the level of CO_2 and as a result the Earth will change solely based on the natural factors that have always regulated the Earth's climate.

The global warming debate will eventually end. The Earth will not stop changing. In the end, humanity will have to adapt to an Earth that does not care where our cities are or where we plant our crops. Eventually the planning that humanity considers will have to take into account periods of time that are far longer than our individual lifetimes. That is one very important point that the warmists and I agree upon.

We should agree on many more, but unfortunately that is not the case. Science should not be like politics. Galileo was a victim of politics when he supported Copernicus against the politically correct view of the day that the Sun rotated around the Earth. Since his view was contrary to what was acceptable, he was told to be silent on the issue. Scientists should never seek to silence others. That this is done today to those that disagree is the saddest part of the global warming debate.

These are extraordinary claims and such claims always require extraordinary evidence. When I reached the conclusion that I did about global warming, I was more than a little surprised myself. That is why I wrote this book, to present the evidence to support my claims. I leave it up to you to reach your own conclusion.

You will find that this book you are about to read contains the history of the Earth's climate and presents the theories about what causes the climate of the Earth to change over time. Most of the theories are already in use by scientists around the world today. What this book does is put all of this information together in a single place.

That is what science is all about. Increasing humanity's knowledge of the world we live in. Using that knowledge and applying it in the real world is what engineers do. Mostly I hope that I have made it simple enough to understand, interesting enough to enjoy and accurate enough to hold up to the inevitable criticism.

In the semiconductor R&D world, one of the most common phrases is "Show me the Data." That is what this book is intended to do. I will show you the data and you are free to agree with my conclusions or not.

Chapter 2   The Earth in Balance

A good place to start is by understanding some basic, but important physical characteristics of the Earth. An advantage of starting with the basics is that it provides an opportunity to show the common ground between myself and the warmists. That in turn will be helpful when disagreements arise. The basics are also very important when it comes to understanding the Earth's climate. For instance, the oceans and the land respond differently to the sun and each has a different effect on the weather and climate.

Weather and climate are different. The easiest way to describe the difference is contained in this saying.

"Climate is what you expect, weather is what you get."

Technically climate is the time averaged behavior for a particular location. That is why Canada expects cold snowy Januaries, but Brazil does not. For many years Canada has had exactly that type of weather so averaging the weather over hundreds of years allows the climate for the month of January to be accurately described. That does not mean *EVERY* January is cold and snowy, but that is the way to bet.

Here are some of the basic physical attributes of the Earth[1].

Radius: 6371 km (mean), 6378.1 km at equator, 6356.8 at the poles.
Total Surface Area = $4\pi\, r^2$ = 510,064,041 km^2

148,940,000 km^2 of land (29.2% of the surface)
361,132,000 km^2 of water (70.8% of the surface)

Aphelion: 152,097,701 km (farthest from the sun in its orbit)

Perihelion: 147,597,887 km (closest to the sun in its orbit)

Then there is the average temperature. Does the Earth have a correct temperature? The commonly used "correct" temperature is 14.0 °C (57.0°F). It is unfortunate that this is not a very useful answer. This can be demonstrated by showing the Earth's temperature in a slightly different way. Sometimes a different perspective changes a persons understanding[2].

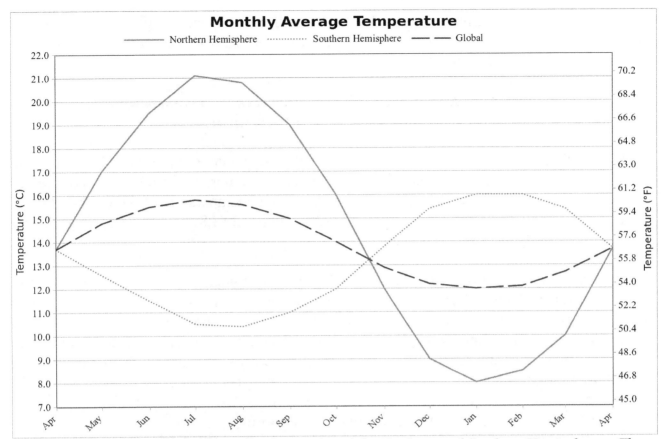

Illustration 1: Annual Temperature of the Earth and the Northern and Southern Hemispheres. The average temperature of the Earth is different for each month of the year.

This graph provides a different perspective to demonstrate that the single value for the Earth's temperature is not a useful description. Knowing the average temperature for a given day is not as useful as knowing how hot or how cold it will get. It also shows that the temperature of the Earth is ALWAYS changing. Depending on the time of year the average "correct" temperature is always going up or going down, much like the temperature changes during the day. The morning is cooler than the afternoon.

Is it surprising that the temperature of the Earth is always changing? Once I saw this it made sense to me, but before I thought about it I would not have thought that the average temperature of the Earth would change so much. If I had been asked about it I probably would have thought that the changing seasons would have balanced the overall temperature so that it generally stayed the same. I was certainly surprised by how much the average temperature of the Earth changes over the course of each year.

In January the average temperature is 12.0 °C (53.6 °F) and in July it is 15.8 °C (60.4 °F). That is almost a 4 °C change that takes place in a period of six months, twice a year. The entire amount of warming that has taken place since 1860 is less than 1 °C. That means that the entire change in the past 150 years is much less than the amount that the temperature of the Earth changes naturally each year. To put this in perspective, the average temperature of the entire

Earth goes up more during the month of March than all of the theorized global warming. It is correct to say that natural global warming happens each March-May and natural global cooling happens each September-November.

Temperature Anomaly: *The difference between the measured temperature and the average historical temperature.*

When temperature anomaly is used (it is the default temperature used in the debate) for a year or for the months of the year, it is showing the difference from normal for that month of the year. So if the anomaly for the month of January is 0.5 °C, then the temperature of the Earth is 12.5 °C. If the anomaly for July is 0.0 °C, the temperature of the Earth is 15.8 °C. So even though January has a greater anomaly, the Earth is still warmer in July.

Here is what the period from 2005-2010 looks like for the temperature and the anomaly. This is also the warmest period that humanity has recorded in the modern era. When a chart is used that gives the appearance of the Earth radically increasing in temperature, it is a chart showing the anomaly data only. The anomaly is the space between the the average and the actual temperature. As you can see that a large anomaly is still a small difference between the actual and average temperature for a given month.

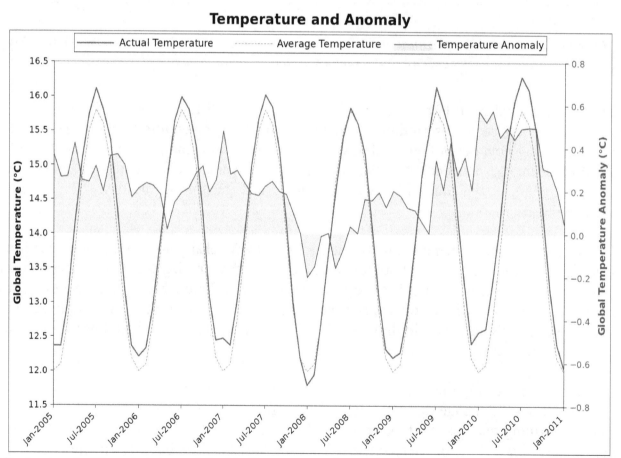

Illustration 2: (Green) Average Monthly Temperature (Brown) Actual Monthly Temperature (Purple) The monthly anomaly.

Understanding this one simple fact puts a different perspective on the entire topic of global warming. As an engineer I also wanted to understand why the Earth's average temperature changed so much over the course of the year. The seasons of the northern and southern hemispheres are balanced. One has winter while the other has summer. There has to be a reason why the seasons in the northern hemisphere seems to have more of an influence on the average temperature than the seasons of the southern hemisphere. Fortunately the answer to that is very simple: the geography of the Earth and the Earth's relationship with the sun.

The seasons on Earth happen because of how the Earth orbits the sun. The main orbital factor that causes seasons is the Earth's tilt. This causes the Northern Hemisphere (NH) and the Southern Hemisphere (SH) to get different amount of sunlight at different times of the year. In December the NH gets much less sunlight than the SH. So the SH has summer while the NH has winter. In June the situation is reversed.

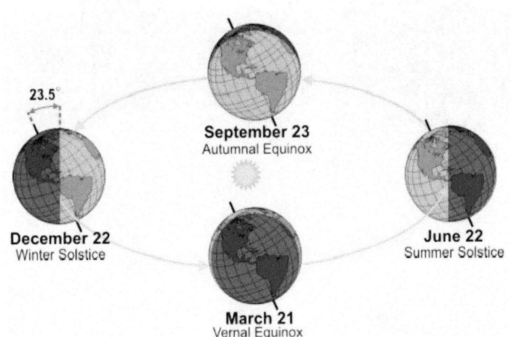

Illustration 3: The Earth's tilt causes the seasons. Picture from NOAA Online School.

The amount of sunlight the Earth gets is the cause of both cycles that most people are familiar with. Every 24 hours the Earth goes through the day/night cycle. The words dawn, noon, twilight are all used to describe the daily relationship that the Earth has with the Sun. There are also words to describe the yearly relationship with the Sun. Spring, Summer, Autumn and Winter are the words to describe the yearly cycle of the Earth. These seasons are directly caused by changes in the amount of sunlight that the Earth receives.

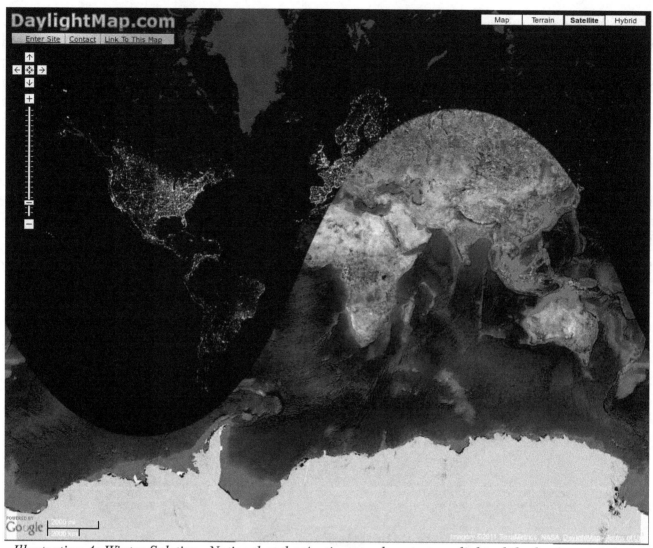

Illustration 4: Winter Solstice. Notice that the Arctic gets almost no sunlight while the Antarctic is in sun 24 hours a day.

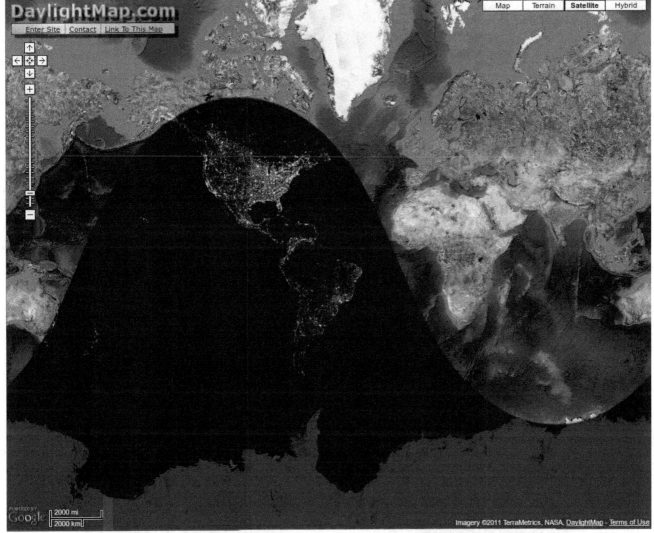

Illustration 5: Summer Solstice. Now the Arctic is in sunlight 24 hours a day while the Antarctic gets none.

This might seem basic and more of what a person learns in elementary school, but there is a reason to refresh your memory. The two basic cycles of the Earth are easy to understand. The Earth has other cycles that are much longer and harder to understand. Keeping these in mind later will help you understand those when I discuss them.

If the NH and the SH were exactly the same, then the average temperature of the Earth would not change much at all. Since the temperature does not stay the same, there must be something about the two hemispheres that is different enough to cause the temperature of the Earth to change with the seasons.

So what is the difference? The answer is that the two hemisphere's have very different geography. Looking at a map is enough to give the correct answer. It has to do with how much land and ocean each hemisphere has.

	Land	Ocean
Northern Hemisphere	39.3	60.7
Southern Hemisphere	19.1	80.9

The NH has more than twice as much land as the SH. Oceans do not change temperature as much as the land does. Since the NH has more land it changes temperature more than the SH. The chart of the yearly temperature shows this behavior. The NH gets both warmer and colder than the SH does. So even though the SH is half of the Earth, the NH is what changes the average temperature of the Earth.

Illustration 6: Much of the Northern Hemisphere is covered by land.

This is true even though the Earth is closest to the sun in January (at least for the next several hundred years). If the land and oceans warmed up the same then January should be the warmest month. Since it is one of the coldest months of the year, this is the first really big indicator that land plays a much larger role than oceans do in changing the average temperature of the Earth.

It is also makes a big difference in WHERE the land is. The SH has very little land around the South Pole. The only large piece of land near the South Pole is Antarctica. That continent is surrounded by water. Since the poles experience the most change in weather from season to season the fact that the South Pole is land surrounded by water is very important.

The NH is the opposite. The North Pole is water surrounded by land. So in the summer the land around the North Pole warms up much more than the ocean that surrounds Antarctica. So the NH warms up more in the summer than the SH. The reverse is true in the winter. The land around the North Pole (Asia, North America, Europe) cools more than the ocean around Antarctica.

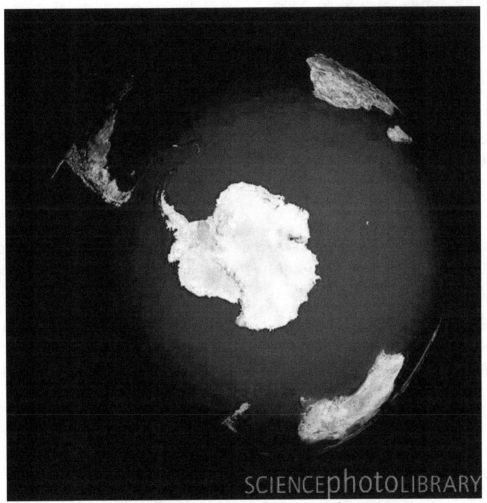

Illustration 7: Southern Hemisphere is mostly open ocean.

That is another part of why the NH has more temperature swing than the SH. It is also why Antarctica remains frozen and also why it has been frozen for such a long time. The water around it does not warm up enough in the summer to cause the air to warm up enough for it to thaw out. While it does warm up by a fair amount it does not warm up nearly as much as the NH does.

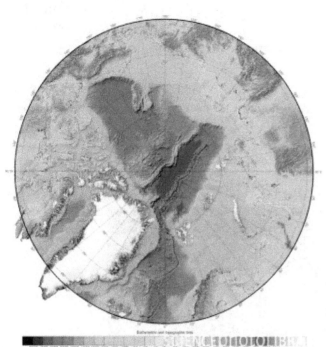

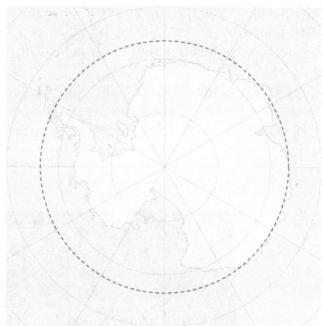

Illustration 8: Antarctic Circle: Mostly land surrounded by ocean.

Illustration 9: Arctic Circle: Mostly ocean surrounded by land.

Greenland is the ONLY land north of the Arctic Circle that doesn't have a large landmass to the south of it. It is no surprise then to find that Greenland is the one place that is the most like Antarctica in its climate. Antarctica and Greenland are the two places where there are large permanent ice sheets. Ice sheets are what glaciers grow into when they continue to grow for hundreds of thousands of years. An ice sheet can cover thousands of square miles and be thick enough to cover mountain ranges. Antarctica has had some amount of permanent ice sheets for more than 34 million years now. The only current ice sheets in the NH are in Greenland and they are less than 3 million years old.

Recognizing that geography plays a role in the Earth's climate might seem like a new idea, but it always has played a role. France and Minnesota are the same latitude, but they have very different climates because they have different geography around them. It is evident that geography plays a powerful role in why the Earth in July is warmer than in January.

The different behavior of the NH and SH is the simplest way to analyze the Earth, but it is the regions closest to each pole that matter the most in affecting the climate of the whole and specifically from about 65 degrees to the poles. I will use 65N and 65S to designate those regions specifically. These are the places that see the biggest temperature change from summer to winter. So when I reference the NH and the SH, remember that I am referring more to the Arctic and Antarctic regions of each hemisphere. The mid-latitudes of each hemisphere are most important

16

in how they affect the Polar Regions.

There is one more important region that needs to be understood. This last region is the preferred place to go on vacation, the warm and balmy tropics.

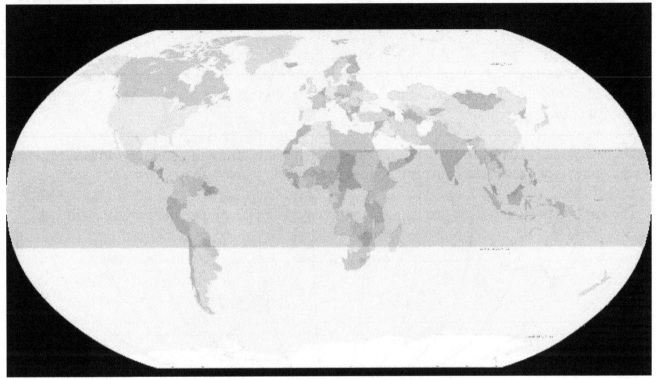
Illustration 10: Tropics are highlighted.

The tropics are important because they are the only region of the world that receives large amounts of excess energy from the Sun. This excess energy is the main driving force for the ocean and atmospheric currents. It is the interaction of the ocean currents and the atmosphere that make the Earth livable for humanity. Without the extra energy that the tropics get there would not be enough evaporation to drive the rains and snows that make the continents habitable. The ocean currents also drive the winds. The currents and the winds are what mix the oxygen into the oceans so that sea life can thrive.

Very little of this should be new information to most people and none of it is controversial. The only thing I have done is presented the information in a different way and explained why the Earth behaves in the manner that it does. Minnesota and France are both the same difference from the equator, but they have different climate because the geography of each place is different. Minnesota is far from the stabilizing effect of the ocean while France has extensive coastal areas that moderate its climate. That same reason explains why Antarctica and Greenland are comparable to each other in climate. Geography ensures that this will remain true for as long as the Earth's geography remains the same.

I did not find this information neatly packaged in the way I have presented it. I found bits and pieces scattered in various places. One sliver of understanding led me to the next question, each layer building up my understanding of the whole picture. More than anything that is what I am presenting to everyone, the entire picture of the Earth's climate. I will only focus on the fundamentals. Not every chapter will be as simple as this one is, but each chapter will build on the one before. In this manner I will lay out a description of the Earth's climate piece by piece so that when the book is done, the reader will have a full understanding of the Earth's climate.

The reason is simple; all of this is tied together. The Earth has been going through natural cycles that have endured for millions and billions of years. It will continue onward for more millions and billions of years. Nothing humanity can do can stop them. These cycles are huge and they will endure.

The one thing humanity must learn is that although the Earth will continue forward regardless of what humanity does, it will not remain the same.

The Earth is in balance, but that balance has always been changing and it always will.

Warning:

Scientific Content!!!

Welcome to the Scientific Content. Each chapter will have a section dedicated to a more in-depth discussion of the content that was presented in the chapter, and more technical details of what I presented in the main chapter. Some of the gentle feedback I have gotten on these sections is that they can be a little boring. They are not required to understand the big picture, but they do fill in some of the details. Feel free to read them or use them as a reference.

Measuring the Earth's temperature is not easy. In a way it is much like measuring the temperature of a person. Where you measure gives different results and what those results indicate can have many different meanings.

If you ask what the correct temperature of a person is, the answer will generally be 37.0 °C (98.6 °F). That is only partially correct. If the surface of your hands are measured, then the temperature will be lower. If you swallow a thermometer, the temperature inside your body will be higher. That 37 °C is generally used to describe the temperature of a thermometer that is put under your tongue.

If a person's temperature increases to 38 °C (100.4°F), then the person is considered to have a fever. If a person's temperature drops below 35 °C (95.0°F), then they probably have hypothermia. A person's temperature can vary a lot without any other cause. Sit in a hot tub for a while and your temperature goes up. Take a walk on a cool evening and your temperature will drop. Your body is rarely exactly 37 °C. On average over the course of the day the temperature of the human body varies by ~0.5 °C.

The Earth's temperature is even harder to measure. The main reason is that it is much, much larger. Measuring its temperature in one place means nothing. Sticking the thermometer into the ground an inch changes the readings. Measuring one meter above the ground gives a different reading than 5 m off the ground. Different methods of measuring the Earth's temperature will generate different results, much like different methods of measuring the temperature of the human body.

Today there are two main methods used to measure the Earth's temperature. The first is what I call station method and the second is called satellite method. Each of them has advantages and disadvantages.

Station Method[3]:

This is the one that is more familiar to people. It involves a thermometer that is 1.5 meter above the the ground and out of direct sunlight. The high temperature and low temperature each day are averaged for each reading to give the daily average temperature. Since land is only a third of the Earth, a similar reading is taken by ships or weather buoys on the sea for the Sea Surface Temperature (SST). The oceans temperatures don't vary much between day and night so one reading is generally enough for the ocean surface.

All the numbers are averaged for a day and that is the daily average temperature for the Earth.

The biggest problem with the station method is that thermometers are not all over the Earth. Some places have lots of them (USA, Europe) and some places have hardly any at all (Africa, Antarctica). In many places a single thermometer is used to measure the temperature for an area larger than Texas. There is no way that a thermometer can accurately do that.

The other problem is where the thermometer is located. If the thermometer is placed above concrete or asphalt, or in a city surrounded by buildings, the temperature for the high and low temperature will be higher than if it was in a field of grass. This is called the Urban Heat Island (UHI) effect. Bricks, stone, concrete, asphalt all absorb heat and retain heat into the night. Being in a city makes the temperature several degrees higher than it would be otherwise. Since many measuring stations are in cities (where the people are), the UHI has an impact on the measurements from cities.

Satellite Method[4]:

This involves a series of satellites in orbit that measure the electromagnetic transmittance of a region of the atmosphere. This is converted into a temperature for that particular region of the atmosphere and averaged over the Earth as a whole. The satellites cover far more of the Earth each day than the stations do, but they don't cover the entire Earth every day, due to the nature of the orbits the satellites are in.

The satellites cover the oceans and areas without stations more completely than the stations can, but they are not measuring temperature in the same manner that a thermometer does. Satellites can also have trouble measuring the lower atmosphere if there are heavy clouds (although they can measure the top temperature of the clouds which the stations cannot measure).

Another complaint against the satellites is that each satellite gives slightly different readings. However, since EVERY station of thousands uses a different thermometer there is a large range of differences inherent in the station method as well.

The satellites overall give better resolution to what is going on with the Earth's temperature than the station method does, but the satellite method only went into effect in 1979 while there is somewhat accurate station data going back into the 1800's. How accurate that early data is can be debated: some of it is very accurate, but for most places on the Earth there is almost no accurate temperature data from before 1900.

Unless specifically stated otherwise, the temperature data I use in this book is a combination of two station methods and two satellite methods. Specifically I use the CRU[5], GHCN[6] station methods and the UAH[7] and RSS[8] satellite methods. In this way I incorporate the modern, more sensitive satellite method with the longer period of the station method. So any data past 1979 includes data from all 4 methods. I call the set I use the blended method in the book and on my website to make it clear that is is a hybrid temperature set. Here they are compared to each other.

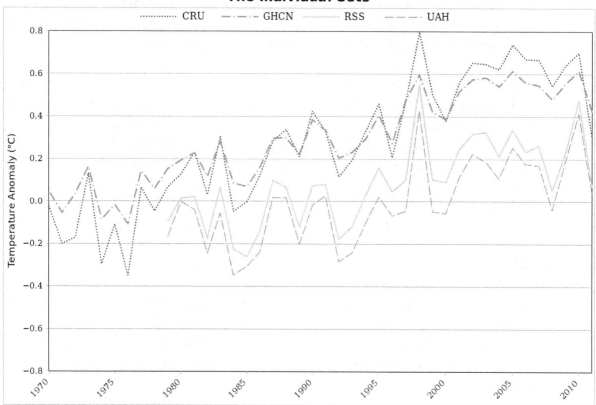

The Individual Sets

.... CRU   —·—·— GHCN   ——— RSS   ——— UAH

Illustration 11: The station methods are generally warmer than the satellite methods.

The two methods that agree the most are the two satellite methods. The two that agree the least are the CRU and UAH, each from a different method. Those two also happen to be the ones that are most used by warmists and skeptics. I will let you guess which group uses which data set.

Since different groups prefer different temperature sets I tried to use all of them as much as possible. There is one other station set that I did not include that is also commonly used and that is the GISS. I wanted to have 2 of each so I chose the ones that I considered the best of each category and have stuck with that. There is also the HadCRUT which is a version that includes the ocean and the CRU (which is land only), but that method interestingly enough matches the blended set fairly well. So instead of using that one, I use the CRU (the oldest one) and the GHCN for the station.

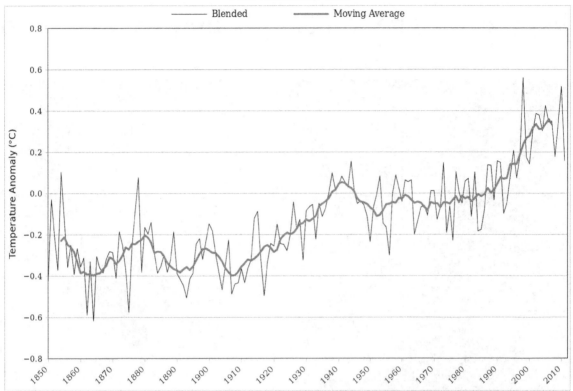

Illustration 12: Using a single set for most usages is useful and incorporates more information than any single set.

The temperatures from the 1800's have a higher error than the more modern data. This is easily seen from the data in 1883-1886. That period is special because August of 1883 is when the eruption of Krakatoa occurred. This caused the the temperature of the Earth to drop by at least 0.5 °C[9]. This lasted for several years, but the CRU and GHCN temperatures from that period do not show the effects of the eruption. Even today the station method shows less sensitivity to events than the satellite method.

The single set that I will use to show the warming that has taken place since 1850 is the above blended set which combines the separate sets. This set shows more warming than the satellite data, but less than the station data. It also shows that the Earth has warmed up by ~0.4 °C since 1980 and I accept that this is an accurate and reasonable measurement, not perfect, but reasonable.

Chapter 3   The Recent History of the Earth

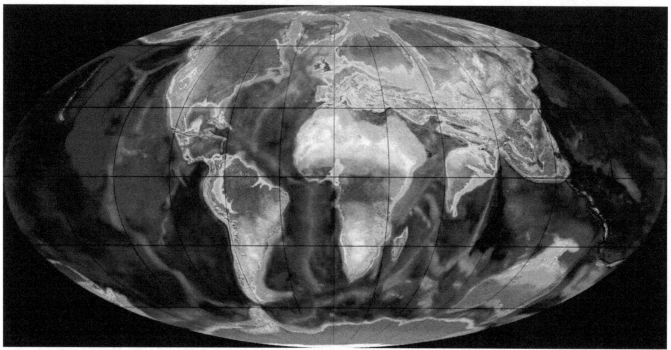

Illustration 13: The Earth 50 Million Years Ago. This is the period known as the early Eocene. Sea levels were more than 150m(500ft) higher than they are today as there were no permanent ice sheets anywhere on the Earth.

During my research I have come into contact with many smart people. The map above was created by Dr. Ron Blakey[10]. He has an entire series of maps that show the evolution of the Earth's geography. I have always liked maps[11] and this one is truly spectacular. I can only show a few of them in this book, but you can be sure I will show as many of them as I can.

I found this map when I got onto the topic of the temperature history of the Earth. That is a natural topic of study when trying to understand why the Earth is the way it is today. In the past the Earth has been both warmer and cooler than it is today by at least 10 °C.

There are many theories out there for why this is, but I have found that the simplest explanation is the best one. Geography[12] is a very important factor to the Earth's temperature. The Earth was warmer 50 million years ago because the geography was different. This can be proven by a very simple example. Suppose that Antarctica and Hawaii switched places. If everything else was exactly the same, the Earth would be radically different than it is today.

First off the sea levels would be ~73m[13] (215 ft) higher than they are today. This is because there would be no large permanent ice sheets at the South Pole. The ice sheets in Antarctica depend on the land for their long-term existence. Without the landmass underneath, only seasonal ice would exist at the South Pole. This one geographical change would result in a much warmer Earth than exists today. On average the Earth would be at least 5-6 °C warmer with that single geographical change, even without considering any other factors that would arise. It is a critical point to understand that a change in geography will result in a change in the Earth's temperature, even if everything else was the same.

Conversely, if the North Pole had identical geography to the South Pole, the Earth would be colder than it is today and the sea levels would be lower. Such a prediction is easy and safe to make because looking back over the past 50+ million years shows how changes in geography affect the Earth's temperature.

Hopefully it is clear why I showed that map at the beginning of the chapter now. It shows the best known reconstruction of the Earth's geography from 50 million years ago. It is not difficult to see how that Earth is related to the modern Earth, but the changes that have happened to the Earth's geography have also caused the Earth to cool over that same period of time.

One of the main factors that cause geography to have such an influence on temperature is the ocean currents. The oceans play a critical role in controlling the Earth's climate because that is how the Earth transports energy from the tropics to the cooler Polar Regions. Anything that alters the flow of energy will have an impact of the climate of the Polar Regions.

As the continents have drifted the ocean currents have changed. It is clear from geological records that dramatic changes to the climate have happened when the ocean currents have been changed by the landmasses either moving together which stops a current from existing, or from pulling apart which allows a new current to form.

This is why it is pointless to directly compare the Earth today to the period 50 million years ago. The ocean currents are different now because the continents have forced them to be different. The Earth is a much colder place today than it was in the past because the Earth itself is a different place.

Geologists are the ones that deal with the history of the Earth's changes the most. They have broken that history down into many different names[14]. The current era is the Cenozoic and it started 65 million years ago. Geologists picked that point because that was when the Chicxulub asteroid hit the Earth and wiped out ~75% of all species on the Earth[15]. That is one of those traumatic events that occasionally happen to the Earth. Just like a car crash can change a person's life, the beginning of this era started with a large and sudden change to the Earth.

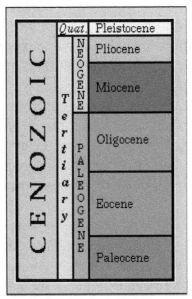

Illustration 14: The epochs of the Cenozoic. UC Berkeley.

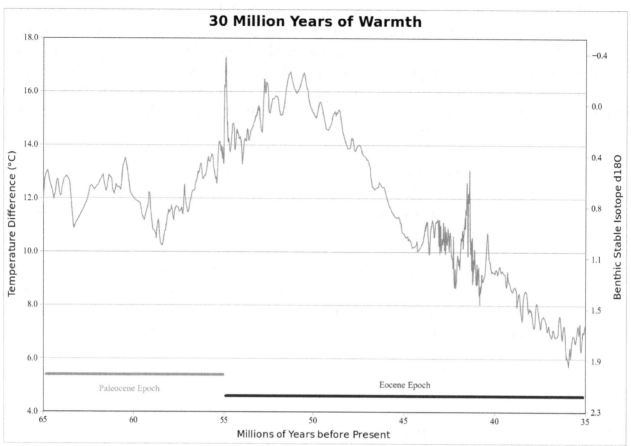

Illustration 15: Temperature was determined by calibrating stable isotope to temperatures of the recent glacials. Epochs of the Cenozoic are also shown. (Zachos, 2008)

In the beginning of the Cenozoic, the Earth was much warmer than it is today. Most of the past 65 million years has been warmer than today. The current Era started off roughly 8-10 °C (18 °F) [16] warmer than it is now. The major difference in temperature was at the poles. Antarctica had no permanent ice and was probably comparable to Canada as we know it today. The Arctic was probably even warmer than Antarctica.

This is most evident from the climate that existed at the beginning of the Cenozoic. When the era started, there were no glaciers or ice caps on the Earth. Both Polar Regions were too warm for those to exist[17]. Today the average Arctic temperature is -14 °C (7 °F) which is warm compared to the average Antarctic temperature which is a stunning -50 °C (-58 °F)[18]. When the era started both regions had average temperatures that were above freezing[19].

The mid-latitude regions between the poles and the tropics were also warmer than they are today, but to a lesser extent than the Polar Regions. It is also worth noting that the tropics were not that much warmer than they are today[20]. So even though the Earth's average temperature has always been changing, the higher the latitude, the greater the change has been.

So when you look at the temperature charts, don't think of the whole Earth being that much warmer, but keep in mind that most of the temperature change has happened in the Polar Region and the smallest temperature change has happened in the tropics.

The reason that the Polar Regions were warmer is the ocean currents[21]. 65 million years ago the ocean currents were able to directly come from the tropics to the poles. The warm water from the tropics helped keep Antarctica and the North Pole warm and prevented permanent ice from developing. There is evidence that the Arctic region was warm enough for at least temperate forms of sea life[22] to live there. It was probably comparable to what the Mediterranean is like today.

Part of this was caused by the fact that Australia and South America were closer to Antarctica at the time. This forced the ocean currents to loop from the equator all the way down to Antarctica. That constant flow of warm water kept Antarctica warmer than it would have been. Much like the Gulf Stream today helps keeps the Atlantic warmer than the Southern Ocean that surrounds Antarctica.

Since there weren't any ice sheets in Antarctica or in Greenland the sea level was much higher than it is today. They were likely more than 150m (500 ft) higher than they are today. A direct comparison is difficult because the continents themselves have changed. Large shallow oceans that existed are gone and others have expanded.

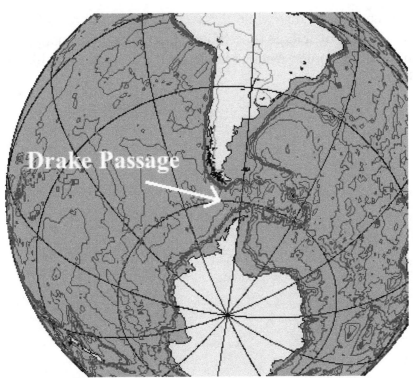

Illustration 16: Drake Passage was closed in the early Cenozoic.
The opening of the passage altered the Earth's climate.

This warmest part of the Cenozoic is called the Eocene Optimum. That is the very warm period that happened 50 million years ago. It was even warmer then, than it was during the time of the dinosaurs. According to the theory of global warming the Earth in that state should have stayed warm. The CO_2 level was high and with many warm, shallow oceans the water vapor in the atmosphere would have also been higher than it is today. It was the perfect condition for a warm world.

That was also the condition from which the world started to cool down. This cooling happened with high temperatures in the Polar Regions and high levels of CO_2 in the atmosphere. Of course it wasn't the entire Earth that was cooling; the real culprit was Antarctica, which started to cool down. 41 million years ago Antarctica was cold enough for snow and ice to form during the winter. This also coincides with the opening of the Drake Passage between Antarctica and South America.

Over the course of 7 million years the passage continued to open. As the passage became larger it allowed the formation of a new ocean current that is today known as the Antarctic Circumpolar Current[23] (ACC). As the ACC grew in strength, less warmth from the tropics reached Antarctica because the ACC is a current that completely surrounds Antarctica. This changing of the ocean currents resulted in the steady average cooling trend that happened from between 34 to 41 million of years ago.

It is important to understand that while the average temperature of the Earth was decreasing, the average drop was caused by a large drop in temperature that was happening in Antarctica. The whole Earth was not getting cooler, Antarctica changed from what was likely a temperate place to a frigid wasteland.

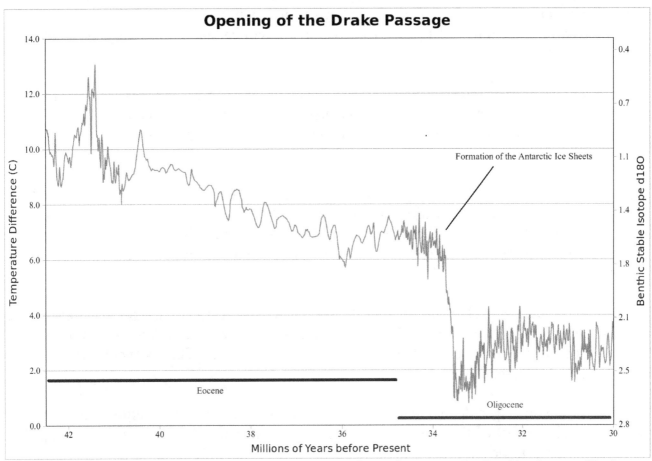

Illustration 17: Changing geography plays a major role in changes to the Earth's climate. (Zachos, 2008)

A little over 34 million years ago the ACC became strong enough to block the warm tropical currents from reaching Antarctica, when that happened Antarctica started to form the permanent ice sheets that exist today. At the time when Antarctica started to freeze over, the world was roughly 4-7 °C (12 °F) warmer than today. The formation of the ice sheets was enough to drop the average temperature of the Earth to only a few °C above what the Earth is now.

This was not caused by a change in the atmosphere, but an alteration of the ocean currents. When Antarctica stopped getting warm water from the tropics it started to freeze. For the past 34 million years Antarctica has remained at least partly frozen. It froze while the average temperature of the Earth was already much warmer than it is today.

Events like this are what cause real climate change. The climate of the Southern Hemisphere was changed for millions of years as a direct result of the geological change of the continents. As the Southern Ocean cooled it also changed the composition of the atmosphere. Large changes in ocean temperatures always cause changes to the atmosphere.

Antarctica will remain much like it is today for as long as the continent itself remains where it is and the ACC remains the same. There have been periods where it has been warmer and cooler than it is today, but the climate itself has been stable and will be stable for a very long time into the future.

The freezing of Antarctica is also a recent example (by the Earth's standard) of the two different hemispheres acting independently. The NH remained warm for tens of millions of years after Antarctica started it's current Ice Age[24]. A change in one polar area does not mean that the other one will follow.

There was one global effect of Antarctica freezing over. That was the drop in the global sea level. The sea levels around the world dropped ~55m as the ice sheets in Antarctica developed 34 million years ago. The effect of the dropping sea level would have been felt around the world. Shallow seas over continental shelves would have disappeared and more land would have been above sea level.

So severe was the global impact that the largest species extinction since the time of the dinosaurs occurred. Species that had been separated by oceans were free to colonize new continents much to the detriment of the slower and weaker species. This extinction is called the Eocene-Oligocene extinction[25].

Sea level change, species extinction and radical change in the Earth's climate all naturally occurred as a side effect of the change that happened when Antarctica's climate changed. The global effect was significant enough that the time before Antarctica froze is called the Eocene epoch of the current era and the time after is called the Oligocene epoch. It was truly a global change that was caused by a long-term change in ocean currents.

This is not to say that Antarctica has stayed the same for the past 34 million years. It has had many periods where glaciers and ice sheets grew, but it has also had many periods when the glaciers and ice sheets were shrinking. That is their natural behavior. For more than 20 million years Antarctica swung back and forth between periods of deep cold and periods that were not as cold, but it never thawed out completely.

That back and forth tug-of-war between warmth and cold stopped for the most part 13 million years ago. The cold won the battle then and it has kept Antarctica in a deep freeze since. If things had stayed the same as they were 13 million years ago the average temperature of the Earth would be 3-4 °C (5-7 °F) warmer than it is now.

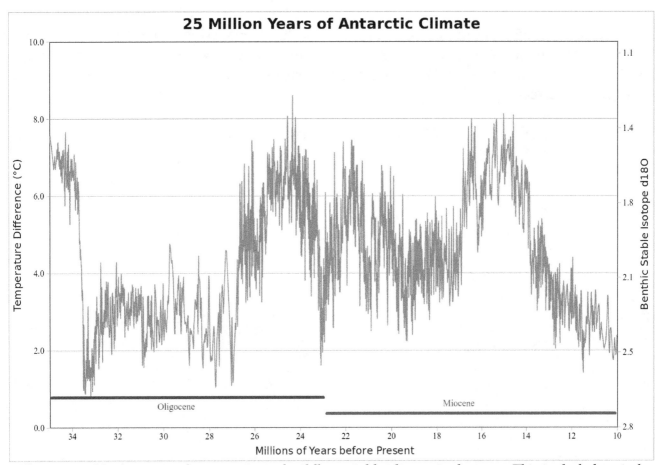

Illustration 18: Antarctica has experienced wildly variable climate in the past. This included periods of rapidly advancing glaciers and also rapidly retreating glaciers. (Zachos, 2008)

Antarctica froze while the Earth was much warmer than it is today. The oceans were higher than they are today, but even with all of that, Antarctica has gotten colder over time. The ice has grown thicker and the life that survived the transition from warm to cold has now adapted to the most brutal climate on the Earth today. The life that did not adapt is now extinct.

The fact that Antarctica froze is the primary cause of the cooling that the Earth has experienced over the past 50 million years. The tropics have experienced very little change in temperature in comparison, but because one part of the Earth got much colder, the average of the whole dropped.

Throughout the entire period of time that Antarctica was cooling and then freezing, the Arctic region in the NH remained warm. The tropical ocean currents continued flowing northward which helped keep the region warm.

As Europe and Asia merged together and India continued to push northward into Asia, the warm ocean currents weakened. As the ice sheets in Antarctica grew, the sea level dropped as well. The combination of factors continued to weaken the warm currents that kept the North Pole warm. Finally 2.6 million years ago it was not enough. For the first time in the Cenozoic the Arctic region started to freeze.

It was at this point 2.6 million years ago that the Earth entered its current Ice Age behavior. While the Antarctic ice sheets that had existed in some form for over 30 million years already, it was 2.6 million years ago that the northern hemisphere started to have periods of glaciers advancing and retreating.

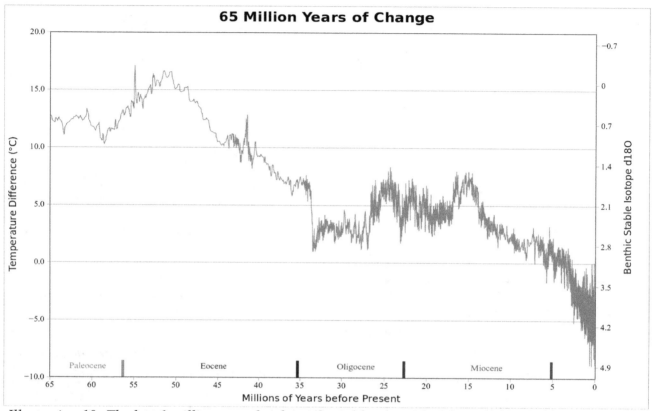

Illustration 19: The last 1 million years has been the coldest period of the past 65 million years. Same temperature calibration to benthic stable isotope. See science content for details. (Zachos, 2008)

No greater example of the independence of the hemispheres is more evident than this period. For 30 million years Antarctica had been in an ice age, but the NH remained more temperate than it is today. The two hemispheres show the record of events that impacted the whole world. Those world-wide changes can influence each hemisphere (by changing the sea level), but they are independent of each other. Of course as Antarctica got colder and the ice sheets grew on the continent, the sea levels around the world dropped. That change in sea levels probably helped contribute to some of the changes that happened later, but still the NH climate remained mostly independent of the changes that had taken place in Antarctica.

Now might be a good time to introduce some definitions so that the proper terms can be used.

Ice Age: *A time of extensive glaciation covering vast areas of the earth.*

Glacial: *Is an interval of time within an ice age that is marked by colder temperatures and glacier advances.*

Interglacial: *Is an interval of time within an ice age that is marked by warmer temperatures and glacier retreats. Marks the periods between glacial periods.*

By these definitions the Earth is currently in an Ice Age, but experiencing an interglacial period between glacials. Geologists all agree that this is the correct state of the Earth at this time. The ice sheets starting to grow in Antarctica was the real start of the current Ice Age, but it was 2.6 million years ago that the northern hemisphere started an ice age[26] with the alternating glacial and interglacial periods.

The glacial cycles that have taken place over the past 2.6 million years have taken place on two different time scales. During the initial period of the ice age, the cycle took ~41,000 years to complete. The temperature range of the Earth during these short 41,000 year cycles was from 0 °C to -6 °C using today as the zero temperature anomaly. The cycles behaved in that manner from the start of the current ice age until 1 million years ago.

For the past 1 million years, the length of the cycle has been ~100,000 years. In addition to taking longer for each cycle to complete, the change in temperature has also become larger. Instead of just a 6 °C change to the average temperature of the Earth, the difference is now 10-12 °C. Over the past million years the average temperature of the Earth has varied from an anomaly that is -11.0 °C to 4 °C warmer than it is now.

Over the past 2.6 million years there have been over 40 glacial/interglacial cycles. Most of these happened when the cycles were short.

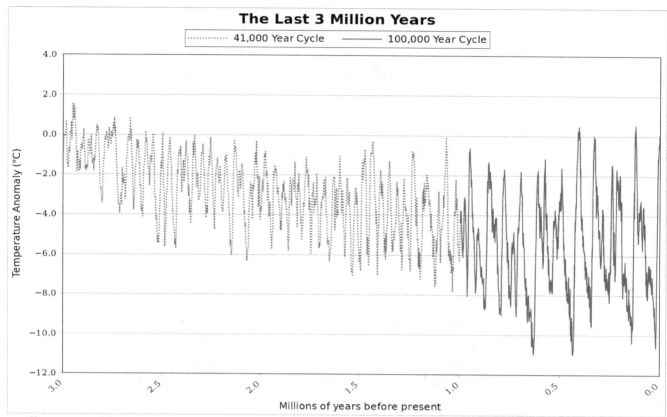

Illustration 20: Temperature history of the last 3 million years. Different source and slightly different calibration. Each time the temperature goes up on this chart is the start of an interglacial. Each time the temperature drops that is the start of a glacial. (Raymo, 2005)(27)

27

For the entire period of the current ice age there has been a steady swing back and forth between warm and cold. The best comparison for a glacial/interglacial cycle is to simply compare each cycle to a single year. The winter would be the glacial period and the interglacial would be the summer. There is also a brief period of transition that could be compared to the spring and fall.

Much like each year is different from other years; each glacial/interglacial cycle is different. They are similar to each other, but they have distinct differences. When comparing the seasons year to year it is the general behavior that matters, not the exact behavior. It is the same when comparing each glacial/interglacial period.

In studying these climate cycles, scientists have found it easier to name the different seasons of these cycles. The name for the current summer (interglacial) is the Holocene and the one before was called the Eemian. These two interglacial periods and the glacial period in-between will be the main items that I will use to explain where the Earth is in these cycles. This combined period of time covers the last 135,000 years. That is the minimum amount of time that should be examined in order to understand what is happening to the Earth today. Any focus on shorter periods of time without an understanding of the last 135,000 years is simply a waste of time.

34

Eemian Temperature Anomaly

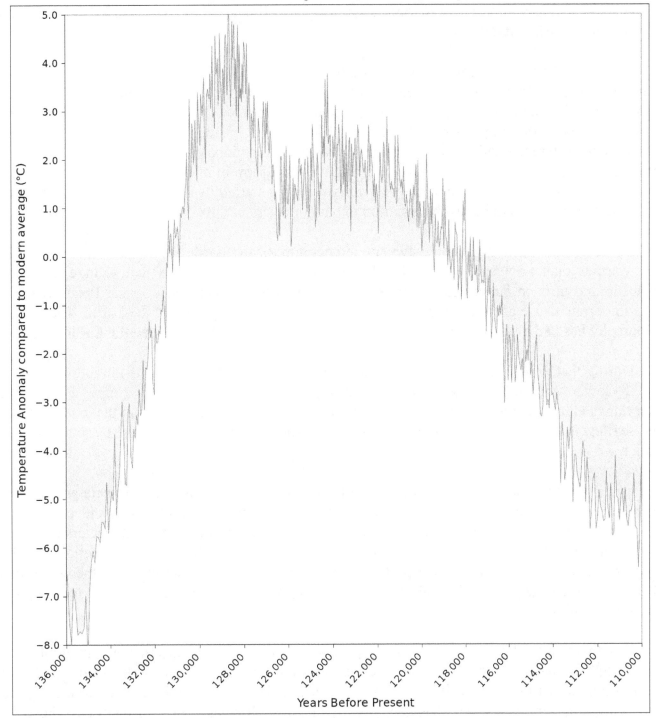

Illustration 21: Eemian temperature reconstruction from the EPICA ice core. This is the temperature reconstruction of the last interglacial the Earth experienced. This is the last summer that the current interglacial should be compared to.

Making predictions of the Earth's future climate without considering the nature of the glacial/interglacial cycle would be like predicting the temperature of January based on the daily temperature trend of a single week in August. No given week of summer weather is useful in making predictions for January. Daily temperature trends are simply not that useful in predicting climate. Predicting climate requires understanding the climate of the past and knowing why it behaved the way that it did.

In the past 65 million years, the one thing the Earth did not experience was a period that was free of change. Over short or long periods of time there is always some kind of change happening. The more detailed the reconstructions, the greater the variation that can be seen. Change has been the only constant for the Earth's climate. In fact, different parts of the world often show different trends in temperature. That makes a global average much less useful in predicting short-term behavior because the Earth is not uniform in its change. A person need only look at the history of Antarctica to see the truth of that. It started its ice age more than 30 million years before the northern hemisphere started to build glaciers in the current ice age.

In the past 65 million years the average temperature of the Earth has been at least 16 °C (29 °F) warmer than now and 10 °C (18 °F) colder than now. Most of that change has been concentrated in even larger changes in temperature that have taken place in the Polar Regions. One important cause of those changes has been the changing geography of the Earth. The altered geography has had a lasting impact on the ocean currents and that has changed the Earth's climate.

During that period of time the oceans have also been at least 150m (500ft) higher and 130m (425ft) lower[28] than they are today. It is safe to say that the Earth changes over time. For anyone that thinks all the change happened millions of years ago, just keep in mind that the lower limit of the sea levels happened a mere 20,000 years ago, a mere blink of time for the Cenozoic Era in which we live.

I am sure that many people could (and will) argue that events 34 million years ago are not relevant to understanding the climate today. My response to that argument would be that the conditions that led to current Ice Age, started with the creation of ice sheets and glaciers that have existed in some form for the last 34 million years. In order to understand the direction that the Earth's climate is headed today, the past must also be considered. That includes the conditions that started the Antarctic Ice Age 34 million years ago and the Northern Hemisphere Ice Age that started 2.6 million years ago.

Now that mankind has the technology to precisely monitor the changes that are taking place in the Earth's climate, it is important that we also expand the scope of our understanding. If our expectation is that the Earth will stay the same as it is today, we are in for a rude awakening. In the past one million years there has been a wide range of climates. Most of them have been far colder than the one we are currently enjoying. Only by increasing our knowledge can we hope to predict what the next phase of the Earth's climate will be.

Warning:

Scientific Content!!!

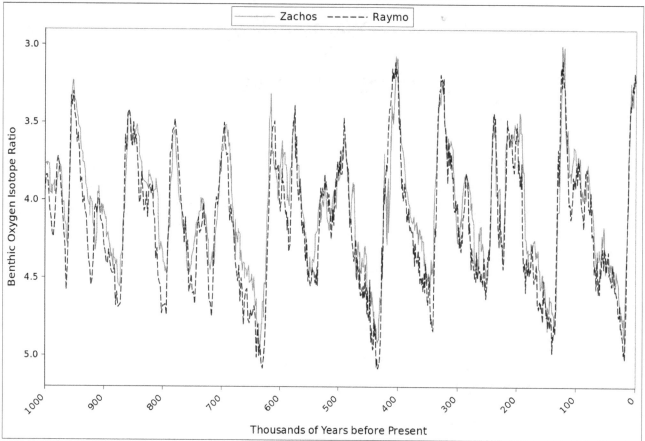

Illustration 22: Two different sources of the benthic stable oxygen isotope data that is used to reconstruct the Earth's temperature for the past 1,000,000 years.

The above chart shows the proxy data from which the Earth's temperature has been reconstructed for the past 1 million years from two different sources. The Zachos paper is the source used for the past 65 million years and the Raymo paper covers the past 5 million years. Since there are no direct temperature measurements for the past 65 million years, a temperature proxy is needed to reconstruct the historical temperature of the Earth. The proxy used for both papers is the ratio of the stable oxygen isotopes.

Stable isotopes of water are used because the ratio of 16 and 18 Oxygen isotopes change with the temperature of the water[29]. What is directly measured is the ratio of the isotopes in an ice core or deep sea sediment record. The sets for both of the million year records are deep-sea sediment cores. The sediment is from the benthic organisms[30] that inhabit the ocean. When the organisms die the isotope ratio records the temperature of the water they lived in. The sediment builds up over millions of years. Scientists can drill cores of this sediment and read them much like an ice core or layers of rock in the ground. It is a very clever and creative solution to determining the historical temperature of the oceans.

This is not a direct temperature measurement, but a reconstructed temperature measurement. Calibrating the isotope ratio to a particular temperature gives some margin for error. Different locations have slightly different behavior. I used a linear behavior for the 65 million year record, in order to simplify the chart. That is also why I put the isotope ratio on the chart as well. Other calibrations have been done that give different temperature results, but I didn't want a very complicated sliding scale chart. So while the absolute scale of the temperature 50 million years ago may not be precise, the relative behavior is correct.

Throughout the book I will put temperatures into the commonly used zero "anomaly" just to keep things consistent. Since all of the data will be available on my website with links to the original data anyone can try different methods to calibrate the isotope records.

I did the calibrations to the maximum and minimum temperatures of the recent glacial / interglacial periods. There have been many papers done on calibrating the stable isotope ratio to temperature and there will be many more. If a perfect calibration is ever done then the precise temperature for a given ratio will be known. The chart does accurately reflect the relative change in ratio and temperature and that is enough. Especially when one considers what a sliding scale the temperature anomaly is anyway.

One of the factors that drive the climate changes over periods of millions of years is the changing geography of the Earth. 50 million years ago there was no Mediterranean Sea. There was a vast ocean from the continent of India to the continental shelf north of the African continent. The portion of the Earth that is now Spain to the Himalayas, were a series of tropical islands surrounded by warm oceans.

All of that land has now merged together and the warm oceans and seas that separated them are gone. Only the small Mediterranean remains. Oceans do not have big temperature changes, land does. As the lands merged together to form a large and more continuous land mass, the NH started to experience more significant swings in temperature. A true tipping point was reached 2.6 million YBP when the balance between the landmasses and ocean currents changed enough to initiate the modern ice age. Eventually what is now Siberia will straddle the North Pole in a similar way that Antarctica covers the South Pole.

When that happens it is also extremely probable that it will be covered in ice sheets that dwarf those of Antarctica. The current ice ages will be much warmer than what will happen in the future. This will not happen for tens of millions of years, but that is the direction the climate of the Earth is going. Nothing humanity can currently do will change that in the slightest.

The changing geography of the Earth is one of those very long-term changes that people cannot see. Those changes do affect the behavior of the Earth's climate. These are gradual changes that take place over millions of years. There is no reason to worry about those changes in the next million years.

There are times when the climate does experience changes that are large and take place over short periods of time.

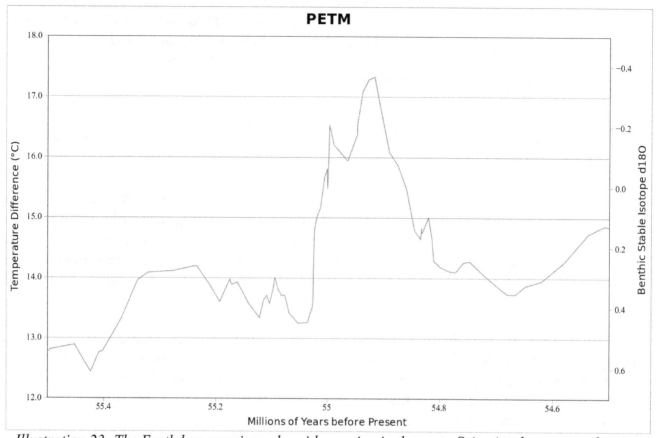

Illustration 23: The Earth has experienced rapid warming in the past. Scientists have many theories for the rapid warming 55 Million YBP.

55 million years ago there was the most recent which is known as the Paleocene-Eocene Thermal Maximum (PETM). In a very short period of time the temperature of the Earth increased rapidly[31]. While generally the chart may slightly over-estimate the temperatures 55 million years ago, it is possible that the spike in temperature at that point was much, much larger than the chart can show. The Arctic Ocean exceeded 22 °C[32] in this period. There are many theories as to what happened, but there are problems with all of them. All that is really known is that the Earth had a period of very rapid warming and for about 200,000 years the Earth was much warmer than at any other time in the past 65 million years.

It isn't a surprise the some warmists have proposed that it was caused by a rapid increase in methane[33] and CO_2. I will not speculate as to what triggered the change, but I bring it up only to point out that the Earth also cooled down almost as rapidly as it warmed up. This is a fine example of the Earth staying in balance. Large changes can quickly take place, but something always balances out those changes. That balancing force will be explained later in the book.

Chapter 4   The Earth's Climate Cycle

Being able to predict the future is something that most people would like to be able to do, from picking the right stock to who will win the Super Bowl. Some of the reasons for wanting to be able to know what the future holds are selfish and some of them are not. Wanting to get rich quickly in the stock market would be nice, but ultimately that is a selfish goal.

For a farmer that depends on the rainfall, knowing the future would be very important. A farmer that knew there would be severe drought that would cause a crop to fail would then be able to either not plant or perhaps grow something different that could handle a drought. In either case he would be able to make the best plan possible for the future.

Since people cannot know the future, they do the next best thing, they try to predict it. Some future predictions are very accurate. No one gets excited at the most basic prediction that the sun will rise tomorrow. The exact moment of sunrise for any location on Earth can be accurately predicted far into the future. Knowing if it will rain at a specific time on a day a month in the future is currently beyond our capability to predict. It might always be beyond our capability, which means that outdoor weddings planned months in advance will always remain in jeopardy.

The Old Farmer's Almanac is an example of trying to predict what the future seasons hold. This almanac has been published every year since 1792. It should be no surprise that the longest continuously published periodical in North America is about predicting the weather. How accurate it has been depends completely on who you ask, but since it has managed to stay in publication for more than 200 years it would seem that people continue to purchase it and use its predictions.

When it comes to predicting climate, things are a little easier. It is a safe prediction that the Northern Hemisphere will be colder in December than it is in June. It is even safe to say that on average the entire NH will be 10 °C warmer in June than December. That is a prediction that will remain accurate for a very long time.

This is because the driving force behind the seasonal climate change is well understood and easy to predict. The Earth's orbit is very well known and so is the tilt of the Earth. From those two factors it is easy to predict that the seasons are not going to change anytime soon. The question that arises in the global warming debate is how the seasons themselves will change in the future. That is what the global warming debate is really about: being able to predict how the seasonal behavior will change in the future.

It isn't really a question of whether it will change because, based on the recent history, it is clear that change is the normal behavior for the planet Earth. It has been changing constantly for the past 65 million years. There is no reason at all to believe that it will stop changing in the future. Expecting it to stay the same would be comparable to predicting that the sun will never rise again.

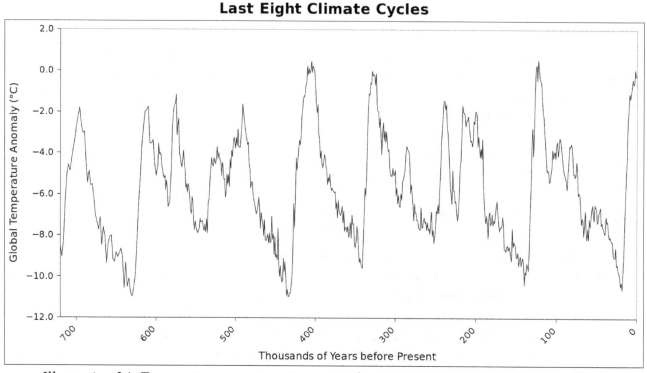

Illustration 24: Temperature reconstruction using deep sea sediment cores. (Raymo, 2005)

With the idea in mind that the Earth's climate will change, it is time to focus on the Earth as it is now. The main focus will primarily be on the last 135,000 years. This will cover two interglacials and one glacial period. Earlier I mentioned that the Earth has longer cycles besides the daily and the yearly ones. That 135,000 year period is the minimum amount of time needed to cover one of the longer cycles that I discussed earlier.

I am going to call the longer cycle of the Earth the Climate Cycle. The climate cycle has seasons just like the yearly cycle does. The hardest part to understanding the climate cycle is simply the time scale that it takes for a cycle to happen. The yearly cycle lasts 365.25 days, or 365.25 times longer than the daily cycle. The last climate cycle lasted about 120,000 times longer than the yearly cycle. Thinking of natural cycles that take this long is a challenge, but that is how the Earth has been behaving for the past million years or so.

The climate cycle does contain more noise than the yearly cycle, but the similarity between the two cycles is obvious. In the yearly cycle, each point is a single month. In the climate cycle each point is 1,000 years. They are two different cycles operating on very different time scales. It is clear that the Earth has just recently (~15,000 years ago) experienced it's 8th significant warming period in the past 720,000 years[34].

Instead of summer in the climate cycle, there is an interglacial. Instead of winter there is a glacial. If I also consider the daily cycle which has a warm part (day) and a cool part (night) then there are at least three different temperature cycles operating together that drive the Earth's climate.

Here is the time scale for the different cycles. I am going to focus on the past interglacial to determine the times for the following chart. This is because there is some variation in the length of the climate cycle. From the start of the last interglacial warming (135,000 years ago) to the most recent (15,000 years ago) the last entire climate cycle on average measures ~100,000 years (the latest one was slightly longer than average).

To help compare the durations of the different cycles I have put together the following table.

	Hours	Days	Years
Daily Cycle	24	1	0.00274
Yearly Cycle	8766	365.25	1
Climate Cycle	876,600,000	36,525,000	100,000

* There has been some variation of the climate cycle over the past 1,000,000 years.

Determining the cause for the two shortest cycles is easy. They are both directly caused by the relationship between the Earth and the Sun. The duration of the two shorter cycles are precisely determined by the time it takes for the Earth to rotate on it's axis and revolve around the Sun. These two cycles have many things in common.

The shortest cycle is the daily one. Dawn is the moment that the sun crosses the horizon and sunlight washes across the land. The day is still cold, but it starts to slowly warm up as the sunlight warms the surface of the Earth. When the sun is straight up in the sky it is called noon. The day is still not at it's warmest, but that is the moment that gets the most energy from the Sun. Things keep warming up until the late afternoon, but when the evening begins, it slowly starts to cool down. Sunset is the moment that the sun disappears below the horizon. The Earth steadily cools down from there until the next day.

The next cycle is the yearly or annual cycle. The best starting point for comparison to the daily cycle is to start with spring. That is the moment that the Earth is evenly facing the sun and is known as the vernal equinox. Spring is the time of transition from the cold weather to the warm one (my apologies to readers in the Southern Hemisphere). It is an unpredictable season with days alternating between warm and cold.

Summer begins with the longest day of the year which happens on the solstice. Much like noon gets the most sunlight in a day, the official start of summer is when one hemisphere gets the most and longest amount of energy from the Sun. Summer is the highlight of the year. That is when the crops are growing, the family is camping at the lake and it is just nice to be outdoors enjoying the warm weather. The long sunny days are the *cause* of the warm summer.

Autumn is like the sunset. It matches the start of spring for the amount of sunlight, but every day is getting less sunlight than the day before. The shorter days and weaker sunlight is the reason why the Northern Hemisphere cools down in the autumn. Mornings are getting cooler and the last season is quickly approaching.

Winter starts on the shortest day of the year. By this time there is snow in many places and even the days are cold. There is sunlight during the day (in most places at least), but it is short and weak. This prevents the Earth's surface from warming up very much and the long nights allow the temperature to drop more and more.

These patterns repeat themselves over and over. These are cycles that people are familiar with in their lives because they happen over and over in our lifetimes. They are predictable because we see them happen. People do not worry that the next day or the next season will not arrive, because they understand the way that the seasons behave.

Notice that within the yearly cycle, the daily cycle continues to behave in the same manner, but the amount of sunlight that a portion of the Earth gets each day depends on the season. In summer, the days are longer and the sunlight is stronger during the day. In winter the reverse is true. The days warm up in the morning all year long, but how much it warms up depends on the season. The night time behavior is also different between the summer and winter. The longer nights allow the surface to lose more energy than the shorter summer nights.

The climate cycle is almost as predictable as the other ones. The problem is that it takes place on a time scale that is closer to the age of humanity than the lifetime of a single person. It cannot be seen in our lifetime and because of that it is harder to observe and people are less aware of it. The evidence for it is obvious, but for the most part it is only the scientists that study it. In addition, it is only in the past 40 years that scientists have really started to understand the climate cycles. In a way, the debate about global warming centers on the exact nature and cause of the climate cycles.

The climate cycle has four distinct phases or seasons much like the daily and yearly cycles have. It can be simplified into two phases much like day and night, warm and cold seasons. The warm phase is the interglacial and the cold phase is the glacial. When most people refer to an ice age they are usually discussing what scientists call a glacial.

To keep the comparisons consistent, I will start with the spring of the climate cycle. The spring is still very, very cold, but the temperature is rising. The ice sheets and glaciers in the Earth's Polar Regions are retreating. The climate spring lasts about 6,000 years. In the period of a climate spring the average temperature of the Earth will change 6 °C (11 °F) or more. In many ways the spring is the most intense period of climate change the Earth ever experiences.

Another way that the climate cycle is comparable to the yearly cycle is the location where the temperature change takes place. The tropical areas around the equator experience very little temperature change. That is why Hawaii is a nice place to visit in December if you live in Minnesota (or Idaho for that matter). A majority of the temperature change in the climate cycle takes place in the same place that the yearly temperature change takes place.

Summer is the warmest part of the climate season and is the part most associated with an interglacial. It lasts about as twice as long as the spring does, but it varies in length much more than a yearly summer does. The longest interglacial in the past million years was the one that was ending 400,000 years ago after lasting ~20,000 years. Climate summers are great times for life on Earth. Forests expand into the areas that were covered with ice sheets and animals expand with the forests. The warmer oceans allow coral reefs to expand and life blooms all over the Earth. The average temperature of the Earth is 10-14 °C (18-25 °F) warmer than the climate winter.

The climate fall can be every bit as unpredictable as the yearly autumn can be. A yearly fall can have brutal winter weather long before the winter officially begins, the climate fall can have abrupt and unpredictable changes. The official fall season will last about as long as the spring, but generally it takes about 5,000 years for the temperature to transition fully into the glacial winter. The glaciers that melted away during the summer start to form once again during the fall. There is no better indicator that fall has arrived than the formation of new glaciers. The farther from the poles that glaciers can be found, the deeper into the climate fall the season is.

Together the first three seasons are referred to as the interglacial. Together they last about 20,000 years in total with the different seasons lasting more or less, but the total being about the same each time. Of those 20,000 years in total, about 12,000-14,000 years can be as warm or warmer than the Earth is today. The remainder is either warming from winter or cooling to the next winter.

That leads us to the truly dominant season of the modern climate cycle, Winter. For the past million years the climate winters (henceforth called glacials) have averaged 80,000 years long, but they have been nearly 100,000 years long at times. The climate winter is a long and difficult season. It has it's own cycles and variations, but in general, for the entire duration the Earth is much colder than it is today. The glaciers that started in the fall slowly grow into continental ice sheets that are miles deep. These ice sheets weigh so much that they cause the continents themselves to sink deeper into the Earth[35].

	Spring	Summer	Fall	Winter
Daily Cycle	Dawn-Noon	Noon-Evening	Evening-Sunset	Sunset-Sunrise
Yearly Cycle	Mar 21st - June 21st	June 21st - Sep 21st	Sep 21st - Dec 21st	Dec 21st - Mar 21st
Climate Cycle	5-7,000 years	8k-20k years	5-7,000 years	80-100,000 years

Duration of each "season" in the different cycles of the Earth.

This has been the climate cycle for the past 1 million years. The climate cycles do change over time, but the trend for the past 2.6 million years has been for increasingly longer and colder climate winters. The interglacials are the brief respites from the brutally long and cold glacials.

The climate cycles are not quite as regular as the yearly cycles, but considering the time scales and the significant changes that happen to the Earth during each cycle that is not a surprise. The last warming period also appears to be late in arriving when compared to the earlier cycles. In both the annual and the climate cycle, there is a difference in how warm each one gets and how cold each one gets, but in both cases the total change in temperature is similar in both of the temperature cycles.

This leads to an interesting question. How should one proceed to predict the future with this information available? When the time scale of the climate cycle is 100,000 years, how significant is 10 years, 100 years or even 1,000 years? The answer is that short periods of time are not very significant. Each piece of data for the climate cycle represents the average temperature for a 1,000 year period.

Any comparison of the temperature trend in the past 100 years is meaningless in comparison to the full scope of the climate cycle. Trends that involve tens of thousands of years cannot be compared to a temperature trend that takes place over 20 years. Such a comparison is absurd in every way.

Let's pretend that the date is Oct 15th and we are brand new to the Earth and settling in Manhattan, Kansas (as I am a graduate of Kansas State University you will have to accept the bias of this choice). When we get there the average temperature is 15 °C (59°F) and the temperature trend for that past 5 days is +0.96 °C/day. Sounds like a great time to plant crops.

Then someone finds evidence that the average monthly temperatures for the past 5 years and they notice that November is colder than October on average and then December is even colder. So there is an upward trend in the short-term temperature record, but the historical climate pattern shows that it should get colder in the future.

What is the correct path to take? Will it get warmer or will it get colder? Some people argue that the past doesn't matter and only the current trend matters. Others are skeptical that a short-term trend matters when the historical records show that on average the future will be colder than the present and in the not too near future it will be cold enough to kill any crops that are planted.

Anyone who has grown up in the Northern Hemisphere will understand that October is not a good time to plant crops. The reason is that it will get colder no matter what the trend has been for a few days. In Kansas, it could be warm far into the Fall, but the temperature can change drastically in a single day.

I will come back to the story of the settlers in Kansas throughout the book. In many ways we are in exactly the same situation as these settlers are. The only substantial difference is that the natural cycle we are trying to grasp is not the yearly one, but the climate cycle. In most respects the only difference in the cycles is the time it takes for the change to take place and how much change actually takes place. In this respect the climate cycle is bigger in every way.

In this story the question of what is normal and what is abnormal is important. Having a warming trend in October is normal. So is having a cooling trend. Both events happen on a regular basis. The only way to determine what is normal is to compare the overall behavior of October to how October has behaved in the past. Any particular temperature trend in the month of October is meaningless. The record high temperature for Manhattan, KS in the month of October is 37 °C (98 °F) and the record low is -11 °C (13°F)[36]. Being abnormal with a potential range like that takes some effort.

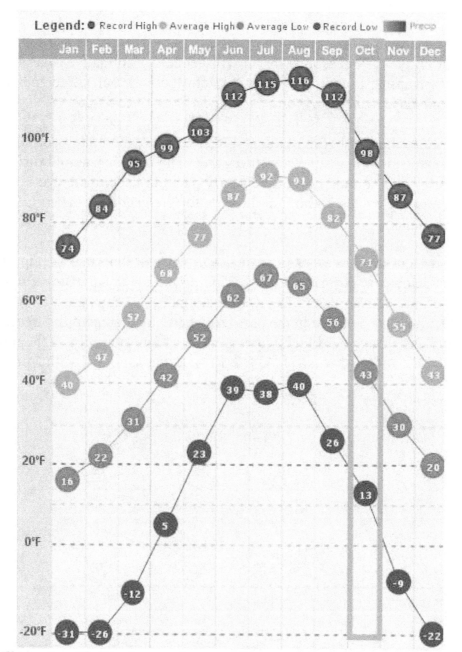

Illustration 25: Weather.com has a useful way of showing how each month behaves over the course of the year. This is their view of Manhattan, Kansas.

For this reason, in order to anticipate the future, instead of looking at the temperature trend for the past 150 years, I am going to show the temperature behavior from the last climate cycle summer. This is directly comparable to looking at the year before to determine what will happen next with the climate.

Just like looking at last year is not going to be a perfect predictor for this year, looking at the last interglacial is not a perfect match for what is happening in this interglacial. It does provide a better predictor for the long-term trends. When it comes to predicting the average temperature for the month of November, it is very likely that looking at last November will be more accurate than

48

projecting from the temperature trend of a few days in October.

The Eemian[37] Interglacial is the name most commonly used to refer to the last climate summer. Looking at the chart below, the last climate summer started 135,000 years ago and lasted until 110,000 years ago. So I will use that 25,000 year period as the basis for understanding how an interglacial behaves. In that way a direct seasonal comparison can be made. In the end the goal is to be able to predict the future behavior of the Earth's climate. Predicting from the average yearly is meaningless unless the long-term behavior is understood. That is why it is important to understand the climate cycle that the Earth has been experiencing for the past million years.

The whole purpose of understanding the climate cycle is to compare the last interglacial to the current interglacial. That is, to compare the Eemian to the Holocene. I did this by using the EPICA ice core data from Antarctica. The results are certainly interesting. In comparison to the last interglacial the current one is much cooler than the last one. The month-to-month comparison also show that the Earth is currently cooler now than it was 17,000 years into the interglacial. There is absolutely no indicator that the Earth is abnormally warm for where it is in the climate cycle.

Interglacial Comparison

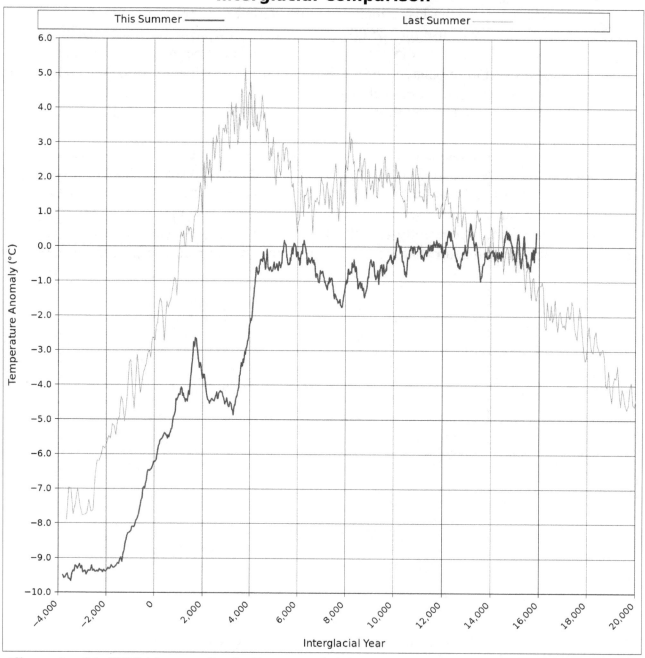

Illustration 26: EPICA ice core for the Eemian and Holocene interglacials. Holocene is this summer and Eemian is last summer.

Warning:

Scientific Content!!!

Within all of the very natural cycles that the Earth goes through, it is very common for the short-term behavior to be opposite of the long-term trend. In the spring and the fall that behavior is even more common. Those two seasons are particularly notorious for rapidly changing weather conditions. Summer weather is usually the most predictable.

In the daily cycle it is very common for weather to alter the normal daily temperature cycle. Clouds that show up early in the day can prevent a day from warming up like it would if clouds had not shown up. People readily accept that each day could bring weather that is not the average. No one panics if the sun comes out in the afternoon and causes warming to happen in the part of the day that is normally cooling.

In the fall and spring the temperature is especially unpredictable. To illustrate this point, here are the day-to-day changes to the average temperature that happened in Manhattan, KS in 2006[38].

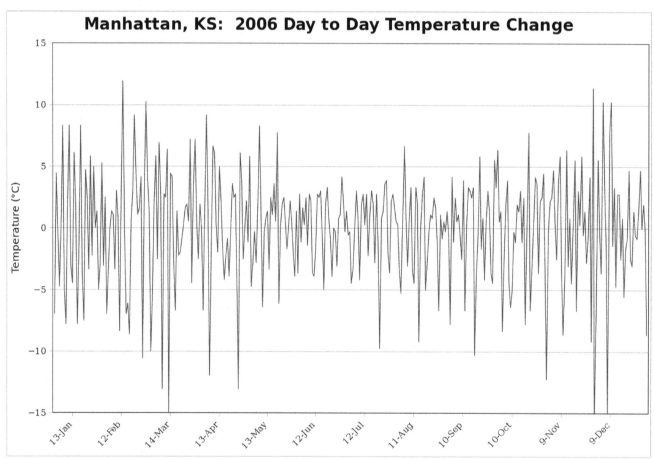

Illustration 27: Throughout the year the temperature can drastically change from day to day. Spring and Autumn experience the most variable temperature.

In most of May-Oct there are very few day-to-day changes that are larger than 5 °C. In the spring they can exceed 10 °C in either direction. That same behavior also happens in the fall. There is more daily temperature variation when the seasons are changing. That is also natural.

This is clear in the amount of variation that happens in the monthly temperature. The standard deviation by month shows the least variation in the warm part of the year and the most in the rest of the year.

Global Monthly Temperature Standard Deviation

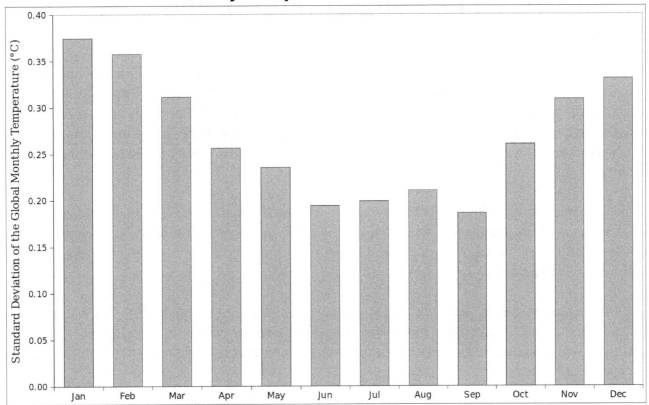

Illustration 28: June-September show the least year to year variation in temperature. December-February are the most unpredictable.

How does this apply to the climate cycle? It shows that the Earth does not behave in an exactly predictable manner. The general behavior for one year applies to the next year, but trying to predict the exact temperature for Oct 15th of any particular year is a guess. A probability could be determined, but betting your house on an exact temperature would be folly.

The climate cycle is even more unpredictable. When a warming period starts, the Earth is in a glacial period. This means that there are vast ice sheets and glaciers that will be melted away. Each spring melt is different in the NH and the same applies to the glacial cycle. All of this makes the climate cycle a more complex and unpredictable one than the yearly cycle.

So would it be a surprise to find that the Earth's temperature can vary a great deal within the large periods of the climate cycle, especially if the Earth was near the spring or fall of the climate cycle? During those periods there is likely to be very strong reversals in the trend that could last for hundreds of years. This is why it is important to understand the climate cycle.

Chapter 5   The Last Climate Summer:
The Eemian Interglacial

Living in Idaho gives my family and I a good opportunity to experience all of the different seasons. Taking advantage of the great weather days is a must. I remember one Sunday in mid-October that was just beautiful. It was a sunny, warm day (22 °C or 71 °F). I knew that the weather forecast was predicting that the very next day the temperature was going to plummet and for at least the next week the weather would be bad, possibly this was going to be the last nice day of the year.

So I decided to take the family to a reservoir for this last perfect day. It hadn't been very cold yet so no one really thought that the summer was over. As a result the kids didn't want to go out on a day that they could sit in front of the TV and play games. What was one more day at the end of a long, warm summer?

I was persistent though and eventually got everyone into the car and off to the lake we went. The kids continued to complain that there was no reason to go the lake that day. It is even possible that I even heard (just once I am sure) that this was stupid. I was patient though and got everyone there in reasonably good cheer.

Jenny set up the picnic while I set up the croquet set. The kids started to run around the park by the lake on what was truly a perfect day. It was just that perfect temperature where running around wasn't too hot, but sitting around and relaxing wasn't cold.

After eating and playing croquet, all kinds of activities started happening. Kids went and played on the small sand beach by the docks. They were building sand castles on a warm, sunny day in October. Kids were wading into the water and splashing each other and that eventually broke down into a game that wasn't tag or hide-n-seek, but some cross between the two.

Jenny and I had a chance to walk around the park together listening to the kids having fun. She was so glad that we had come to the lake that day. Everyone had a great time and as the sun started to go down we packed everything up and headed home. The kids might not have realized it, but they had made the most of the last day of summer.

That day stands out because I knew it would be the last summer-time like day of the year. It isn't often that a day at the lake is enjoyable that late in year. The very next day fall arrived with a chill wind. It got stormy and the weather never again was that mild until the next spring. In Idaho that was probably 6 months away (I went back and checked, the perfect day was Oct 18, 2009 and it was April 15, 2010 before it got that warm again)[39].

I know it is hard to think about something that happened 130,000 years ago in the same way as the end of last summer, but when the topic is climate, that is the only time scale that matters. So whenever the Eemian Interglacial comes up, think of it like comparing two different summers that you have experienced. Some summers have only a few really hot days and some summers it feels like every day is too hot. Comparing the Eemian Interglacial to the current interglacial is no different than comparing last summer to this summer. It would be nice if winter never came, but in the end it always does.

The Eemian Interglacial was one of the warmer interglacials in the last couple of million years. It was not the longest, but it was a much warmer interglacial than the current one. The part of the Earth that had the most warming was the Northern Hemisphere, especially the Arctic regions[40]. This is how all interglacials have been for the past 2.6 million years and that is simply how the climate cycles behave.

One of the best measures for comparing interglacials to each other is to find how far northward the forests manage to grow. Scientists call the northernmost regions that trees can grow the Arctic Tree Line[41]. Regions to the north cannot grow trees because of a combination of extreme cold and permafrost. The extreme cold freezes the sap inside the trees during the winter and the permafrost prevents trees from putting roots into the ground.

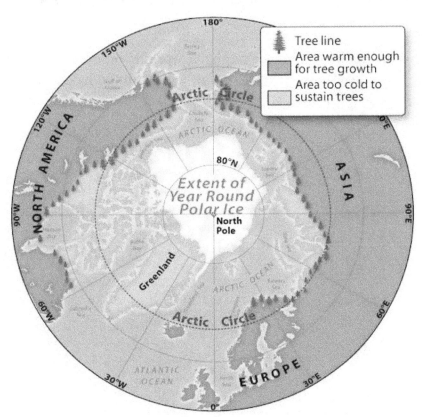

Illustration 29: Modern Day Tree Line. Primarily it was only Greenland that was north of the tree line during the Eemian. Source: Map The M Factory Smithsonian Institution

The tree line is a good indicator because it takes centuries for a forest to develop. As the Earth started to warm up 135,000 years ago it allowed the forests a chance to grow northward. When the peak warm period was reached 128,000 years ago the forests appear to have reached to the northernmost reaches of both North America and Asia[42]. The modern tree line is hundreds of kilometers further south today in most places.

The name Eemian[43] comes from the river Eem in the Netherlands. In the 1870's Dr. Peter Harting found large numbers of fossils of warm water sea life in the soil at various locations in the Netherlands. The strange thing was that the shells were from a warmer climate and didn't match the type of life that was currently found in the Northern Sea. They didn't know exactly how long before it had been warmer in the Netherlands, but the findings were in soil that was only 17m deep, indicating that it was in the recent past.

Modern science has gone back and extensively studied the Netherlands and found the entire area was warmer and filled with many forms of life that cannot live in the present climate. Geologists have mapped the extensive change in the area and there are now many parts of that country that were underwater during the Eemian Interglacial.

*I should note that I will be using the conventional **YBP (Years before Present(Y2K))**. I will always use the oldest time from the left going towards the most recent time to the right of the graph. There is a surprising amount of variation in the literature, but I will keep it consistent in this book.*

So how warm was the Eemian? Science has been working on the issue. Antarctica has several ice cores that show the temperature changes during the last interglacial. Below are the ice cores from the EPICA[44] and Vostok[45] ice cores.

Eemian Temperature Anomaly

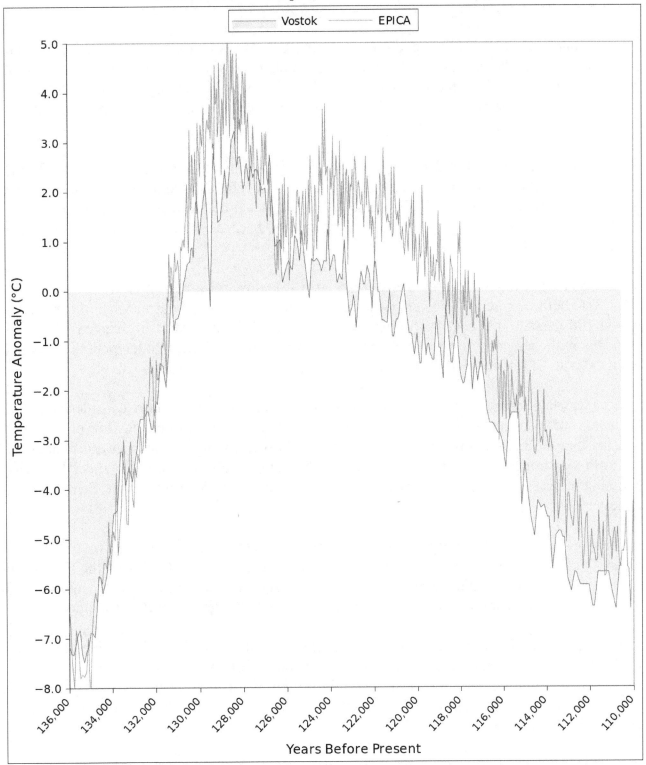

Illustration 30: The EPICA and Vostok ice core temperature reconstructions for the Eemian.

According to the data from the two ice cores, the Eemian was significantly warmer than the Earth is today. Of particular interest to those that believe in global warming, the Eemian was more than 3 °C warmer than the Earth is now for thousands of years. The warmest period of the Eemian is exactly what the most dire predictions of global warming are predicting for the Earth in the future due to CO_2 levels.

So understanding what the climate of the Eemian was like would provide a better understanding of the horrific world that the warmists claim will happen if the projected warming does happen in the next hundred years and beyond. This is a real advantage because the Earth has experienced many difference climates. Reasonable predictions can be made simply by looking at those times when the climate was different.

Another interesting fact is that the Earth was 3-5 °C warmer when CO_2 levels were 270-280 ppm. The same level that the Earth has had for the past several thousand years while it has been much cooler than it was during the Eemian. If the CO_2 level determines the global temperature, then the Earth was broken during the Eemian interglacial because the current theory doesn't work for the warmth of the Eemian.

During the Eemian, the Earth was warmer than it is now for a period of 13,000 years, from 118K-131K before present. The very peak temperature was ~5.5 °C warmer than the average of the past 100 years. That peak warm period took place ~128,700 years ago happened when the CO_2 levels in the atmosphere were ~285 ppm. According to the IPCC, that temperature should be associated with CO_2 levels that are around ~1100 ppm[46].

Why was the Eemian so much warmer than the Earth is today, especially considering that the CO_2 levels were so much lower then, than they are now? That is only the first of two major problems that the Eemian gives to the theory that CO_2 is a driving factor in global temperature. The Earth was much warmer 128,000 years ago, but CO_2 levels were nothing special. Perhaps something other than CO_2 levels caused the Earth to be much warmer during the Eemian besides CO_2 levels?

For now, let me tell the story of a warmer Earth. This story is not a prediction of what may happen, but a story of what once was. There is no better way to understand what the Earth would be like if global warming did happen, than to take a closer look at the Eemian. According to the IPCC estimates, the Eemian is what the Earth will be like if CO_2 levels reach 1,100 ppm.

I will not lie; the Earth was much different during the warmest part of the Eemian. In no place was this more evident than in the Northern Hemisphere (NH). There were dramatic differences throughout the entire hemisphere, especially the farther north one looks. Perhaps the most surprising find were the fossils of many tropical African animals that lived as far north as Germany during the Eemian.

The most surprising of these (to me) was the hippopotamus.

Illustration 31: Hippopotamus today is only found in Africa.

The Hippo can only live in warm climates. An important factor in this is water temperature. Hippopotami need to use water to stay cool[47]. The hippopotami bones that have been found in Germany that date to 125,000 years ago are a very good indication of how warm it was throughout the year in that part of the world. Winters would have had to have been much warmer then than they are today, meaning that Germany was much more tropical during the Eemian interglacial.

The arctic regions, North America, Asia and Europe all experienced much more of a temperature change than the tropics did. That is simply another area where the climate cycle and the yearly temperature cycle are the same. The Northern Hemisphere experiences the most temperature change on a yearly basis and in the climate cycle. The expanded range of the hippopotamus is a good example of how great that change can be.

Illustration 32: The Water Buffalo today is mostly in India and Pakistan.

Another odd animal found in Germany is the Water Buffalo. These two animals cannot survive significant frost, much less extended freezes. That their remains have been found in Germany from the Eemian period is an indication that the winters in that area were much warmer than they are today. Average winter temperatures must have been above 0 °C (32 °F) for them to have survived there for extended periods of time.

It is clear from the fossils in the Netherlands that the seas around Europe were much warmer than they are today. Germany was warm enough for the hippopotamus and water buffalo to survive. There are indications that the elephant and other African animals were able to survive in that region, another good indicator of how much warmer the Eemian period was in Europe.

For people worried about the polar bears, the Eemian period is especially interesting. The oldest fossils for the polar bear date to sometime during the Eemian period or early in the glacial that followed. DNA tests[48] show that the modern Grizzly bear was the parent species for the polar bear and the two animals can still have fertile offspring. Since the two are so specialized in their habitats they are considered separate species. It is considered probable that a group of grizzly bears became isolated with the rapid onset of the glacial as the Eemian period ended. This group of bears then evolved to the current species we all know as the polar bear. It is even possible that the polar bear exists today because a group of grizzly bears became isolated on one of the Arctic islands as the sea ice melted completely for several thousand years. Such isolation would have

60

given the polar bear the start at favoring the skills in swimming that gives it such an advantage today.

One advantage in looking at the longer term for climate is that it is clear that the early Holocene was warmer. When I show some comparisons to the current interglacial it will be clear that the polar bears are very capable of surviving a wide variety of climates. In fact it was the big changes in climate that appears to have given polar bears the chance to exist in the first place.

Illustration 33: Polar Bears as a species began during or at the end of the Eemian. Courtesy of the USGS.

For all the changes that happened on land, it is the oceans that experienced the largest change during the Eemian. It appears that the average ocean temperatures are comparable to the current temperatures, but the warmer waters extended much farther north. One significant impact of this is that much more of the Greenland ice sheets melted during the Eemain than have melted in modern times. The oceans were likely to have been at least 10m (33ft) higher than the modern sea level.

Illustration 34: Momma bear and cub. Courtesy of the USGS.

Compare this much higher sea level to the estimated 1 meter change in sea level that is projected for a CO_2 level of 600ppm. In the past, the seas where higher when the CO_2 levels were less than 300ppm. This is especially a problematic because the CO_2 levels for the past several thousand years were identical to the levels during the Eemian. Yet the Earth during the Eemian was much warmer and the sea levels were much higher.

It is very easy to prove that the oceans were higher during the Eemian than they are today. This is because the coral reefs that lived during the Eemian are now above sea level. These fossil coral reefs can be seen on many of the islands in the Pacific Ocean. Coral reefs can be a powerful tool in understanding how ocean temperatures change, but any corals that lived during the Eemian died when the oceans started to drop when the Eemian ended 115,000 years ago.

The biggest problem that the Eemian presents to the theory of global warming is that it ended. 120,000 years ago the Earth was warmer than it is today by a degree or so. CO_2 levels were ~270 ppm at that point. By 115,000 YBP the temperature of the Earth had dropped 4 °C, but CO_2 levels were still ~270ppm. The Earth had shown dramatic cooling in a 5,000 year period while CO_2 levels remained almost identical to what they had been.

Eemian Temperature Anomaly

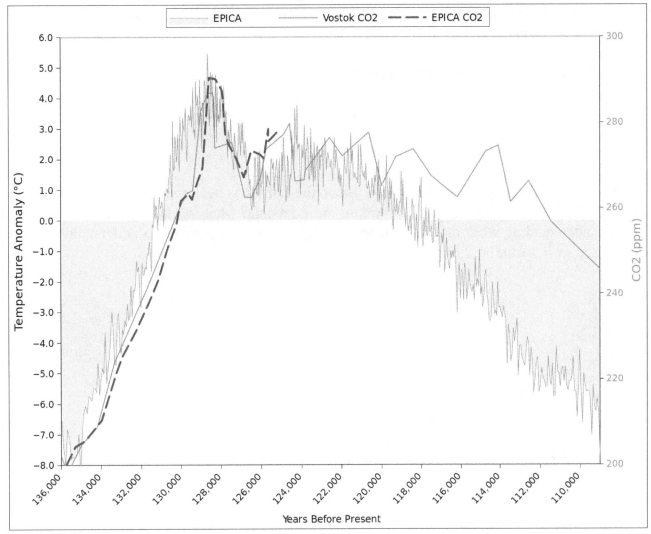

*Illustration 35: CO₂ levels dropped ~8,000 years **AFTER** the temperature dropped. The cooling could not have been caused by dropping levels of CO₂.*

In all, the global temperature dropped more than 10 °C, while CO_2 levels remained constant at ~270ppm. If the greenhouse effect of CO_2 is so strong, why did the CO_2 level during the Eemian fail to keep the interglacial from ending? Since the CO_2 level did not drop until long after the temperature dropped, some other factor must have caused the Earth to cool.

How is it possible to correlate CO_2 levels to temperature when the range of temperature for given CO_2 levels is so broad? That is the only reason why CO_2 is associated with temperature change in the past, but the problem is that it only shows a match to temperature change when the Earth is warming. When the Earth is cooling, CO_2 stays at the same level while the temperature drops. It is only after the Earth has cooled that CO_2 levels start to drop. This should make scientists wonder just how much influence CO_2 has on global temperatures.

In addition, there is the question of what caused the Earth to warm up in the first place? The average temperature of the Earth increased 13 °C at the beginning of the Eemian and that warming started while the CO_2 levels were very low. In both the interglacial spring and fall the temperature change happened before the CO_2 level changed. This is especially evident when the Earth cooled down at the end of the Eemian.

This leaves two basic questions that need to be answered.

What caused the Eemian interglacial?

What caused it to end?

The answer to these questions is critical. Since the Earth has been repeating the same cycle for hundreds of thousands of years, it is important to understand what should naturally happen next. Only by understanding what should be happening right now is it possible to understand if what is happening now is natural or not.

Warning:

Scientific Content!!!

One of the problems with science is that each scientist will name an event long before it is understood where the event fits in the big picture. There is no clearer evidence of this than the naming of geological epochs and the glacial-interglacial periods in the past. The geological time scale is useful, but it is so full of assumptions, that it creates a bias without intending to.

The current glacial / interglacial behavior is the perfect example of this problem. The Eemian does not exist according to the geological naming system. All the previous cycles are stuffed into a single epoch by the name of Pleistocene. Then the current interglacial is given an epoch all to itself by the name of the Holocene. So the current interglacial is treated in a very different way, even in how it is named.

This introduces a bias that this current interglacial is unique, but it is not. It is only from the perception of a human lifetime that the Earth is stable. Geologists are not immune from that bias and certainly in the past there was good reason to suspect that the Holocene was unique, but that is no longer the case.

In addition not everyone uses the Eemian to refer to the last interglacial. Geologists from different continents have named them differently. There are at least 5 different names that are in use to describe the same thing. That makes clarifying the information about the period all the more difficult.

Naming confusion aside, the proxy data is not confused. It is very clear from many different types of proxy data that the Eemian was warmer than the Earth has been during the modern interglacial and the sea levels were also higher during the Eemian.

The main methods of reconstructing past temperatures are deep sea cores that measure the oxygen isotope ratios and also ice cores. [49]There are two methods of determining the past temperature from an ice core. One is to measure the ratio of heavy and light oxygen atoms in the ice and the other is to measure the ratios of heavy and light hydrogen. Since these are the two components of water there are plenty of both to measure in ice. The ratio of heavy and light oxygen is the standard measurement.

Water that is made of the light oxygen evaporates more easily. Water made of the heavy oxygen condenses more easily. This means that the warmer the oceans are near the location of the glaciers or ice sheets, the more heavy oxygen there is in the ice core.

Another way to explain it is how warm the ocean is near the location where the ice core is from. The less heavy oxygen there is, the colder the water is near the glacier. The warmer the water is near the location of the ice core, the higher the content of heavy water there will be in the core.

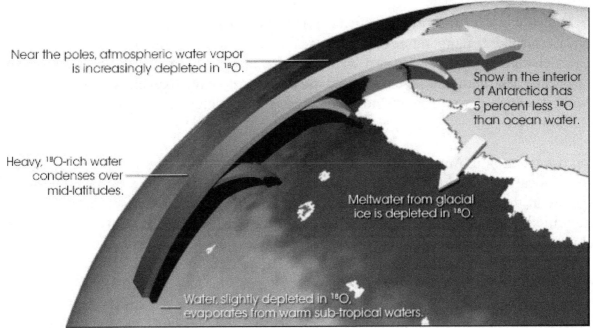

Near the poles, atmospheric water vapor is increasingly depleted in ^{18}O.

Heavy, ^{18}O-rich water condenses over mid-latitudes.

Snow in the interior of Antarctica has 5 percent less ^{18}O than ocean water.

Meltwater from glacial ice is depleted in ^{18}O.

Water, slightly depleted in ^{18}O, evaporates from warm sub-tropical waters.

Illustration 36: Heavy Oxygen Cycle: NASA Earth Observatory

The water that evaporates and falls as snow on a glacier[50] (or ice sheet, but I will only use glacier now) records how warm or cold the water near it was. Then the layer the year after that tells the story of that year. Each layer is recorded one on top of the other. Scientists then measure each layer and count down how many layers there are from today. Then they can know the story told for an exact year. It is like counting the rings on a tree, but it is actually much more accurate. It also tells a much larger story than a tree (or forest) possibly could.

Since Greenland is near the end of the path for the Gulf Stream and most of the water vapor in the atmosphere in that region is from the Gulf Stream, ice cores from Greenland are a good indicator of the ocean temperatures in the North Atlantic Ocean. Ice cores in Alaska tell the story of the North Pacific Ocean. Different regions of Antarctica tell different stories based on the weather patterns of that region.

The simplest reason that it works this way is that cold water does not evaporate very much. Water that is 25 °C (77°F) evaporates about twice as much water vapor as water that is 15 °C (59°F). Since there is much more light oxygen than heavy oxygen, cold water evaporates very little heavy oxygen. The warmer the water, the more heavy oxygen is released.

So ice cores are really telling a larger story than they are often given credit for. That is why they are so useful in understanding the global climate. A coral reef can tell a story, but only at the exact location that the coral exists. Since many of the coral reefs are near the equator, they see less of the change in the Earth's climate and so they tell a very small portion of the story. This is especially true for the past couple of million years as most of the climate changes have been far stronger in the Northern Hemisphere than they have in any other place on Earth.

This is also why ice cores in Antarctica can measure changes that are mostly happening in the Northern Hemisphere. Temperatures in Antarctica do not change as much as the ice cores indicate. What does change is how much warm water is close to Antarctica. During a glacial period (ice age) the oceans near both poles are colder, so the amount of heavy oxygen is very small. When the northern hemisphere is warmer (like now) the oceans have a higher sea level and warmer water is closer to both poles.

Even if a location in Antarctica stayed exactly the same temperature for 100,000 years, the ice core at that location would tell the temperature record of the ocean that evaporated the water that fell as snow at that location. In this way ice cores do not reflect the temperature of the location they are drilled. Ice cores primarily tell the record of the ocean the snow evaporated from and how far that water vapor traveled.

Any type of record that involves the ratio of oxygen that has fallen as rain can also tell the same story. Water that drips in caves to form stalagmites and stalactites can also be used to determine information like this. In places where there are no glaciers this type of thing can be done, but it is more complicated.

Ice cores give the broadest temperature reconstruction because of how the record accumulates. Each layer is distinct and can provide a wide view of the climate for the region for that specific year. This information is recorded for the period that matches the age of the glacier. The bottom and older layers do get squeezed by the weight above, but reliable ice cores that are hundreds of thousands of year old have been recovered.

Chapter 6   Milankovitch Cycles

I would like to jump back to the settlers in Manhattan, Kansas. They have been presented with two different temperature trends. The short-term one is obviously a warming trend. The other trend is based on older records that some argue are no longer relevant because the current trend is warming. They really need to decide if now is the time to plant their crops or if they should wait for some time in the future when it will be safe to plant.

The debate continues on for a few hours and then everyone realizes that what they really need to know is what causes the warming trends to happen. If they knew what caused the long-term warming and cooling, then they would be able to make the correct decision. At this point the two sides start looking at the available information that has nothing to do with temperature.

After some very fast scientific analysis, two theories are presented to the settlers as a whole.

Theory #1:

This idea is presented by a chemist. He knows that gases in the atmosphere can absorb energy. He noticed that the CO_2 level had been strongly increasing for the past few months. Since that level was going up and the temperature was going up it seemed pretty obvious that the increasing CO_2 level was the cause of the warming trend in mid-October. He even put together a chart[51] that showed how much the CO_2 level was going up. The chart showed in an easy to follow manner that the CO_2 level had gone up more than 5% in the last 4 months. That was the obvious reason that the temperature was going up because CO_2 was causing it to warm up.

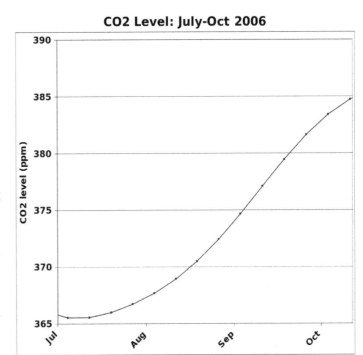

Illustration 37: CO_2 level as measured in the mid-western United States in 2006.

Needless to say people were impressed that he had solved the problem so very quickly. Since he was a well respected chemist, most people were willing to trust that his theory was correct. The temperature trend appeared to match what was going on with the temperature trend.

Theory #2

This idea was presented by an engineer. What he noticed was that each day was getting shorter. As the days got shorter he realized that the area would get less and less energy from the sun. He checked the length of the days against the old temperature records and sure enough, the temperature was warmest when the days were longest and coldest when the days were shortest. Since the days were already less than 12 hours long, the days would continue to get colder. Maybe not today or the next day, but he estimated that exactly one month in the future the area would get 31% less energy from the sun than it was getting now.

He also had a chart[52] to show what would happen to the amount of energy the area would receive over the next month.

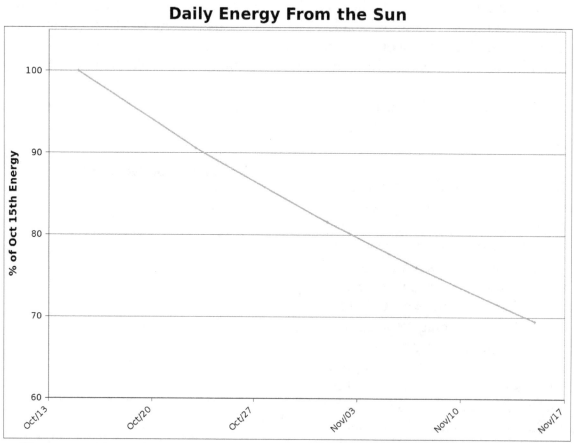

Illustration 38: Each Autumn, the energy drops slightly each day. Sometimes the daily temperature trend (short-term) goes against the seasonal (long-term) trend.

Since this matched up very well with the temperature cycle that happened each year, it was obvious that even though the short-term trend was warming, he realized that colder weather would happen, even if it didn't happen right away.

The supporters of Theory #1 laughed and said that his chart proved they were correct because the temperature was going up, not down. If there was less energy from the sun each day, then the temperature would be going down. Since it was warming, that proved absolutely that the CO_2 theory was correct.

Conclusion:

To be determined....

The above charts are both real. The yearly CO_2 level in the United States for 2006 and the change in energy from the sun are correct. Every year the CO_2 level in the United States has a minimum in July and quickly rises until November.

The change in energy from the sun is also correct for the Northern Hemisphere, although that isn't specific to the year 2006. In both cases there is an argument to be made to support the theory. It is also true that Theory #2 cannot explain why warming happens at times when it predicts cooling. Of course the Earth is also substantially cooler in November even though the global CO_2 level is higher than it is in June.

The real story is more complex, but not substantively different. It was a Nobel Prize winning chemist by the name of Svante Arrhenius that proposed the basics of Theory #1. He proposed in the early 1900's that CO_2 levels were responsible for the past ice ages (they had yet to call it the glacial/interglacial cycle). His idea is the foundation for the modern theory of global warming.

The long-term equivalent of Theory #2 was presented by a Serbian engineer by the name of Milutin Milankovic. His theory was that long-term changes in the Earth's orbit changed the amount of energy that the Earth received from the sun. Specifically he theorized that the long-term changes in solar energy at 65N latitude during the summer months caused what I am referring to as the climate cycle. He presented his idea on the verge of WWII in a book titled: _Canon of Insolation of the Earth and Its Application to the Problem of the Ice Ages_.

So far I have used the term sunlight to refer to the energy the Earth gets from the sun. I am going to switch to the proper scientific term now which is insolation. This is a term that will show up frequently for the rest of the book so it is important to understand. The best way to understand insolation is to compare it to a solar panel. A perfect solar panel that is 1 meter by 1 meter (slightly more than a square yard) that is getting 400 W/m^2 will be able to power 4 x100 Watt light bulbs. Thinking of insolation in terms of powering light bulbs will hopefully make it a little easier to get used to the term.

The yearly seasons are determined by the insolation that each hemisphere gets from the sun, much in the same way that the daily cycle is determined by the hourly insolation. Here is the hourly insolation for the longest and shortest days of the year. This will help explain the **cause** of the daily and yearly cycles of the Earth. In the NH June 21st gets about 3.5 times more total energy than Dec 21st. That is why there is a temperature difference between winter and summer in the NH.

Insolation

Understanding insolation is important. If you think in terms of light bulbs, then what Milankovic proposed was that changes in the orbit caused long-term changes in the summer and winter insolation over periods of hundreds of thousands of years.

If the summer insolation gives 400 W/m² (4 light bulbs), then the winter will typically give 180 W/ m² (less than 2 light bulbs). That is why winter is colder. Less light bulbs of power from the sun.

Modern Day

Summer Insolation

Modern Day

Winter Insolation

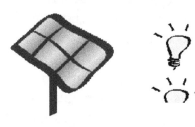

Illustration 1: Summer with 400 W/m² can power 4 light bulbs.

Illustration 2: Winter with 180 W/m² can power only 1.8 light bulbs.

The long-term orbital changes cause the winter and summer insolation to change over time. So if the summer right now is 400 W/m², 5,000 years from now the summer could be 380 W/m² while winter could be 200 W/m². In this case the total energy stays the same, but the winter would be warmer and the summer colder.

**Future
Summer Insolation**

**Future
Winter Insolation**

Illustration 3: Future summer with 380W/ m² can power 3.8 light bulbs.

Illustration 4: Future winter with 200 W/m² can now power 2 light bulbs.

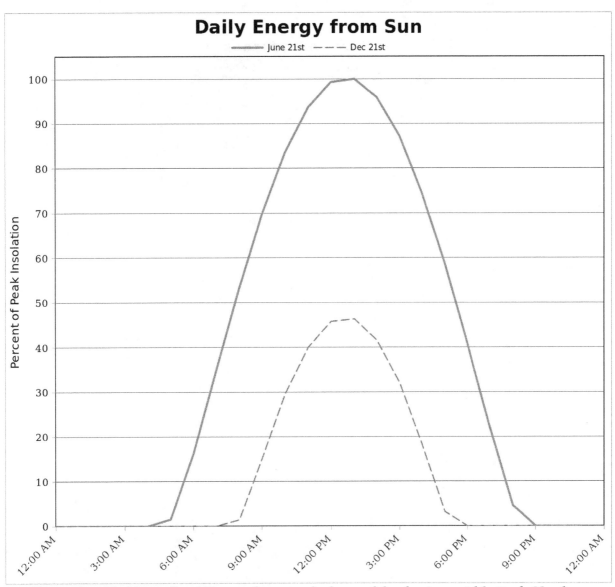

Illustration 39: Normalized Energy from sun by hour of the day in a mid-latitude Northern Hemisphere location.

Winter is colder in the NH because it gets less energy for a shorter period of time. Summer gets more energy for a longer period of time. This energy difference is the entire and complete cause for cold winters, warm summers, chilly mornings and warm afternoons. It entirely **causes** all aspects of the daily and yearly cycles of the Earth.

The seasons behave in the same way. The temperature of Manhattan, KS from 2006 can be compared to the energy it gets over the course of a year. As the insolation increases, the temperature follows. Temperature change is **caused** by the changes in insolation. Insolation changes as a result of the Earth's orbit around the Sun. In the spring when insolation is increasing, temperature is warming up more slowly than insolation increases. The reverse happens in the fall while the insolation is dropping.

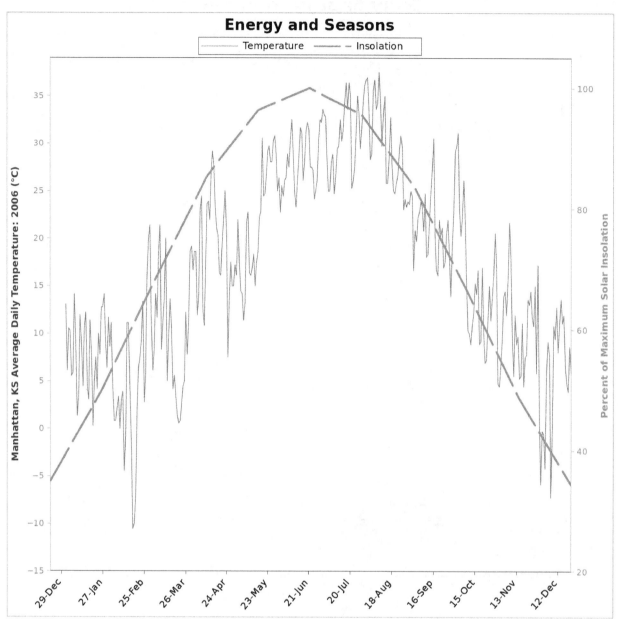

Illustration 40: Annual Energy and the temperature response. The yearly changes to the insolation proceed the changes in temperature throughout the year.

Consider the example of a cold pot of water that is placed on a stove. When the stove is turned on, the water starts to warm up. It takes a while for the pot of water to start to boil. Even though the amount of energy is constant, it takes time for the temperature to respond to the energy being added.

What is happening to the Earth is similar, but instead the stove's setting is always slowly being changed. Anyone who has cooked knows that it takes a while for a pot to reach the desired temperature. If the setting for the stove was constantly changing, so would the pot's temperature. That is the situation that the Earth is in. The stove's setting is constantly being changed. For that reason the Earth's temperature is always playing catch-up to the energy from the Sun.

This lack of immediate response is called time lag or response time. I mention this only because there are many times that the response is delayed in the Earth's climate. The changes in energy are gradual and as a result the Earth does not instantly respond. When looking at the yearly cycle the temperature is behind the energy by about one month.

The time it takes for the temperature to respond is why August is warmer than April. Even though both months get similar amounts of energy from the Sun, in one case the Earth is still warming up and in the other case it is just starting to cool down. The end result is that the NH is 7 °C warmer in August than it is in April. In April the NH is rapidly warming and in August it is just starting to cool, even if it doesn't always feel that way.

Not only is the NH warmer, the Earth as a whole is 2 °C warmer in August than it is in April. That is because the NH experiences more temperature change than the SH. The cause of this difference goes back to the differences in geography between the two hemispheres. The NH has more land. It especially has more land in that critical 65N latitude region that sees the most difference in energy between the winter and summer.

If the hemispheres responded in the same manner, then the warmth in one would cancel out the cold in the other and the average temperature of the Earth would have little variation over the course of the year. This proves the idea that the hemisphere's respond differently to the amount of energy that each gets from the Sun.

Once again, geography plays an important role in the Earth's climate even today. The Earth changes temperature primarily based on the seasons of the Northern Hemisphere. If the Winter, Spring, Summer and Fall seasons are compared by hemisphere, the dominance of the NH is apparent.

This is a very important point about the Earth's current behavior.

The same behavior applies to the climate cycle

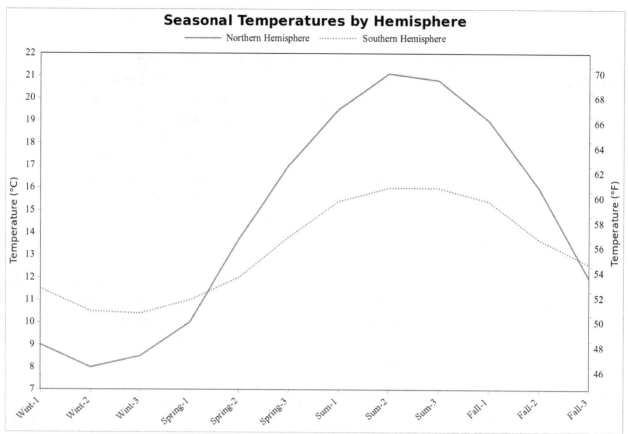

Illustration 41: The NH warms up more during the summer than the SH and it also cools down more.

Prior to the 1970's it was known that there had been Ice Age cycles in the past. The precise timing of those cycles was not known. Since the timing was not known, no theory could be proven true or false. These two theories have been at odds for a long time and it appears that they will continue to be at odds for the foreseeable future.

Theory #2 did get a tremendous boost in 1976. That is when the paper *Variations in the Earth's Orbit: Pacemaker of the Ice Ages* by Hays, Imbrie and Shackleton almost ended the debate for good. They used deep sea sediment cores to measure a sequential and dated historical temperature of the Earth. What they found was that Milutin Milankovic's[53] work on the past insolation of the Earth matched up with the historical temperature of the Earth. It is for this reason that Milankovic's Theory led to what are now called the Milankovitch Cycles[54]. These cycles of energy from the sun cause the climate cycle that I have described thus far.

Illustration 42: Milutin Milankovic
(1879-1958)

The Serbian Engineer Milutin Milankovic is best known for developing one of the most significant theories relating Earth motions and long-term climate change. Born in 1879 in the rural village of Dalj (then part of the Austro-Hungarian Empire, today located in Croatia), Milankovic attended the Vienna Institute of Technology and graduated in 1904 with a doctorate in technical sciences. After a brief stint as the chief engineer for a construction company, he accepted a faculty position in applied mathematics at the University of Belgrade in 1909—a position he held for the remainder of his life.

Milankovic dedicated his career to developing a mathematical theory of climate based on the seasonal and latitudinal variations of solar radiation received by the Earth. Now known as the Milankovitch Theory, it states that as the Earth travels through space around the sun, cyclical variations in three elements of Earth-sun geometry combine to produce variations in the amount of solar energy that reaches Earth.

Here is the insolation anomaly for the NH for the past 150,000 years[55].

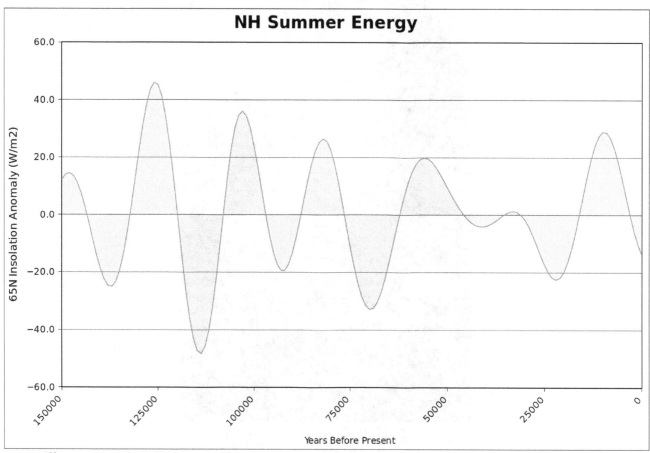

Illustration 43: 65N Summer Insolation Anomaly for the past 150,000 years. (Berger, 1992)

There is a detailed description of the long-term orbital variations that cause the variation located in the science content section of this chapter. In basic terms it is similar to the yearly cycle. The tilt is still a major factor, but the day of year that the Earth is closest to the sun plays a big factor. When the Earth is closest to the Sun on June 21st, the NH summers are very warm. When the closest day is Dec 21st, the summers are less warm. Currently January 5th is the day that the Earth is closest to the Sun. Here is the summary of the cycles and what causes each cycle to exist.

	Duration	Years
Daily Cycle	1 day	Earth's Rotation (Spin)
Yearly Cycle	1 year	Earth's Tilt as it orbits the Sun.
Milankovitch Cycle Climate Cycle	100,000 years	Long-term variation in the Earth's orbit that change the amount of insolation that 65N gets during the summer.

Climate on the Earth is determined by the relationship with the sun. Science has advanced into many other methods to measure the long-term historical temperature of the Earth which include: sea corals, ice cores, boreholes and others. All of these methods paint a picture of the historical temperature of the Earth. All of these indicate that the Sun dominates the Earth's climate.

When the NH summers are cooler for thousands of years, it allows the glaciers to grow larger and larger. Eventually they merge into the massive ice sheets that cover continents. As more of the Earth's water is stored as ice on the continents of the NH, the sea levels drop dramatically. The combination of growing ice sheets and decreasing annual temperatures in the NH combined with the global sea level drop are the defining signatures of a glacial period. All of this is caused by periods of lower than average insolation in the NH summer.

The average energy that the 65N (Polar Region) gets over millions of years is 440 W/m^2. The size of the energy change from high to low is a large percentage of that total energy. From the minimum to the maximum, the difference is 22% of the total or almost 100 W/ m^2 variation over a period of tens of thousands of years. When the summer energy is at it's highest, the Arctic region warms up and occasionally forms an interglacial. When the energy drops, the Arctic cools down and enters a glacial. In this regard, the climate seasons behave in the same manner as the yearly seasons. The energy from the sun and the geography of the Earth dictate the seasons for both of these cycles.

Now would be a good time to compare the Eemian temperature anomaly to the Eemian insolation anomaly. The theory proposed by Milankovic would predict that the Earth would warm when the insolation increased and cool when the insolation did, but in both cases the temperature would change AFTER the insolation did.

Here is the temperature anomaly compared to the insolation anomaly.

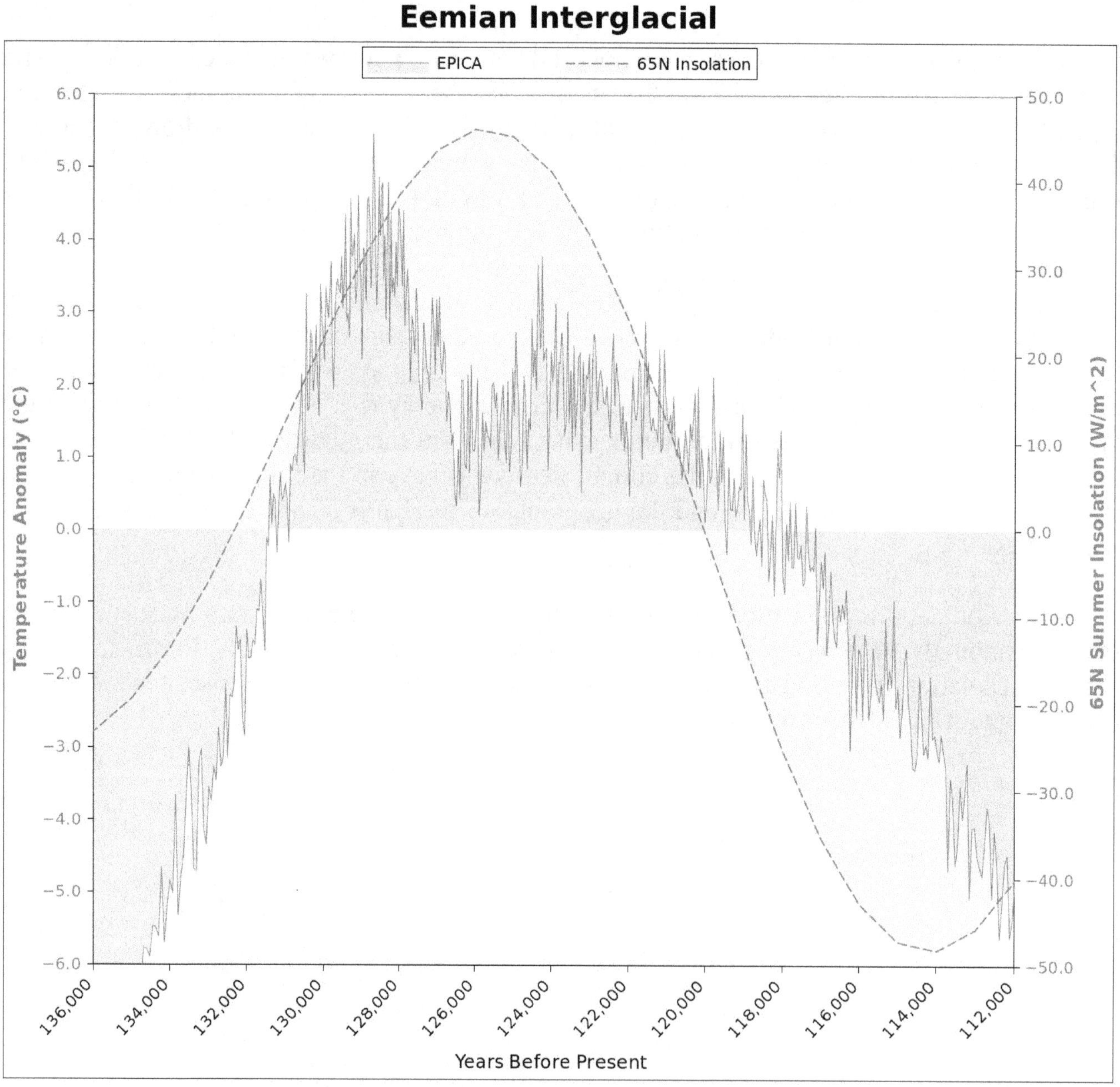

Eemian Interglacial

Illustration 44: The 65N Insolation decreased prior to the drop in temperature. This is exactly the type of behavior seen over the course of the yearly cycle.

This shows that when the insolation was increasing, the Earth's temperature was also increasing. More importantly, it is clear that when the insolation started to drop, so did the Earth's temperature. No other factor can explain why interglacials come to an end. Once the Earth is warm, it should stay warm, but because the insolation in the NH starts to drop during the summer, the Arctic region of the Earth starts to cool down, which eventually leads to a glacial.

The effect of the Arctic cooling is global primarily because of the oceans. The oceans affect every continent and when a glaciation starts in the NH, the global sea levels start to drop. At the end of the Eemian, the sea levels dropped more than 20 m (66ft) in a period of 5,000 years after the 65N summer insolation anomaly went negative 120,000 years ago. The glaciers in the NH were growing so fast 120,000 years ago that the world's oceans dropped as much as a six story building. That is what I consider genuine climate change.

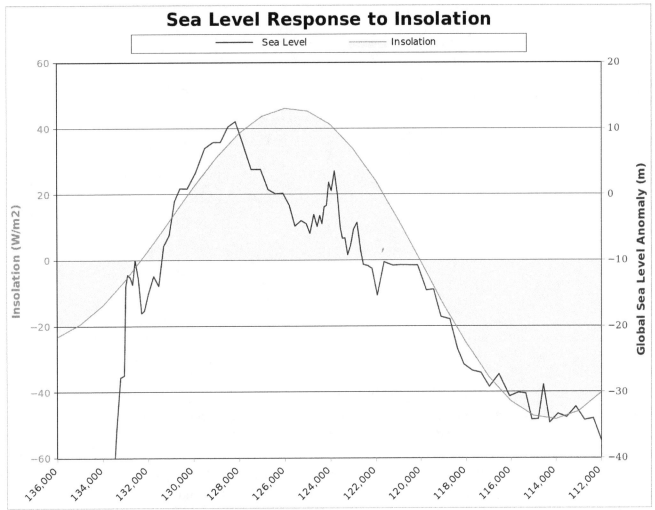

Illustration 45: The Eemian experienced rapid sea level rise when it began and it experienced significant sea level drop when it ended.

In both the climate autumn and the yearly autumn the NH starts to accumulate snow and ice. The difference is that more and more ice accumulates year after year for thousands of years when the NH summer insolation starts to drop. The more ice that accumulates, the colder the climate of the NH gets. This is the onset of a glacial. This is the beginning of the climate winter.

Of course there are two theories for the cause of the glacial / interglacial cycle. Even though the Milankovitch Cycles fit the timing of the start and stop of the climate cycle, it is possible that the CO_2 theory also fits. It can even be argued that it does fit, but only for half of the climate cycle.

81

Eemian Insolation and CO2

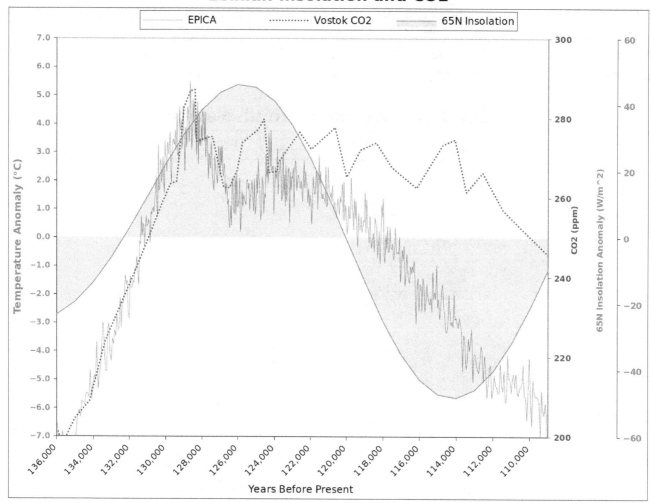

Illustration 46: Insolation and not CO₂ level was the cause of the Eemian cooling. The insolation preceded the warming and the cooling.

After the Earth started to warm up, the CO_2 level also started to increase. While it is interesting that the CO_2 level changed after the temperature did, what I find more interesting is what happened when the Eemian started to dramatically cool down 118,000 years ago. The CO_2 level stayed the same. The Earth cooled down while the CO_2 level was at the "warm" level of 270 ppm.

In fact the CO_2 level stayed nearly constant at 270 ppm from 128,000 to 112,000 years ago. In that same period of time the global temperature dropped 8 °C (14 °F). So while CO_2 did go up after the temperature went up, the CO_2 level didn't do anything to stop the Earth from cooling down. So it could be argued that CO_2 "might" have contributed during the beginning of the Eemian, but the Earth cooled down as the insolation dropped.

It is also interesting to note that the temperature of the Earth during the Eemian was warmer than it is today, while the CO_2 level was only 285 ppm. A temperature projection based on the CO_2 level during the Eemian would indicate that the temperature anomaly of the Earth today should be 1.6 °C warmer than it was at the peak of the Eemian which would put the modern temperature anomaly at 5.4 °C. That is ~5 °C higher than the current temperature.

If the modern insolation anomaly of -13 W/m² is used to predict the modern temperature anomaly, then the prediction would be 0.6 °C which is almost exactly what the modern temperature is. In fact the Earth is 0.2-0.4 °C cooler than the Eemian prediction of temperature. So the insolation gives a very good prediction of the modern temperature while the CO_2 level is a terrible predictor of modern temperature.

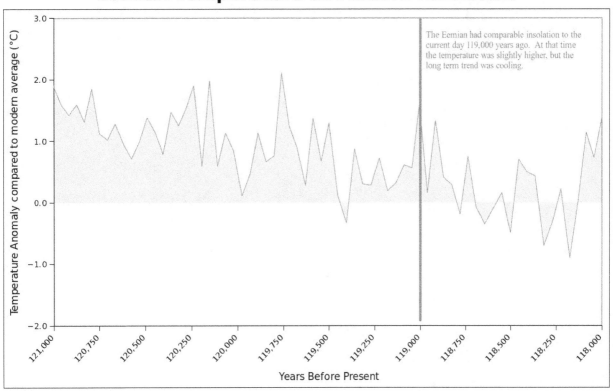

Illustration 47: EPICA Anomaly was comparable to current anomaly. The 2,000 year period had a more than 3 °C range in that period of time. That is much more than the current variation that is causing worry about Global Warming. Temp range on current warming is typically +/- 0.5C

Whatever role CO_2 might have played during the warming part of the Eemian, it played no role in the Earth's temperature as the Eemian ended and the Earth cooled towards to one of its coldest glacials of the current ice age.

Warning:

Scientific Content!!!

The Longest Interglacial

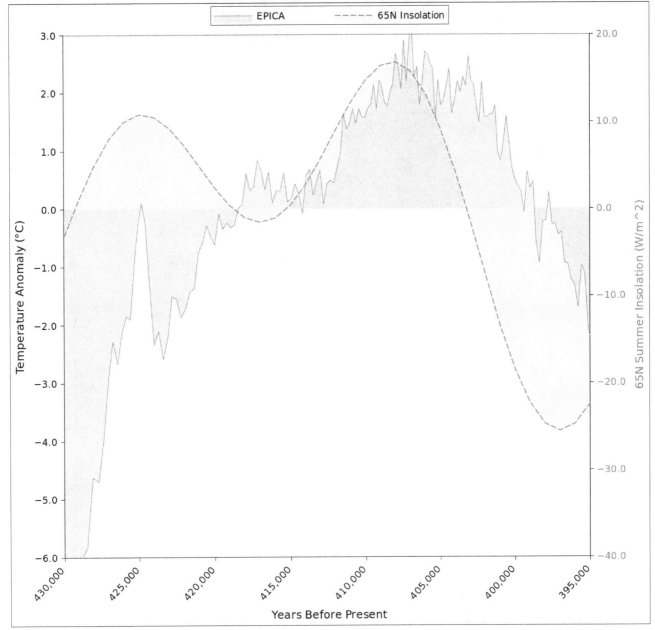

Illustration 48: Longest and most unusual of the current Era of interglacials. Insolation barely dropped below 440 W/m² before increasing again. That kept the interglacial going much longer than normal.

This interglacial was the longest warm period the Earth has had in millions of years. Instead of the normal large drop in insolation it had two weaker peaks, but a very shallow drop that barely dropped below 440 W/m². That meant a long period without any low NH insolation. The Earth responded with temperatures that were as warm as today for ~20,000 years.

Typical Interglacial: 330,000 Years Ago

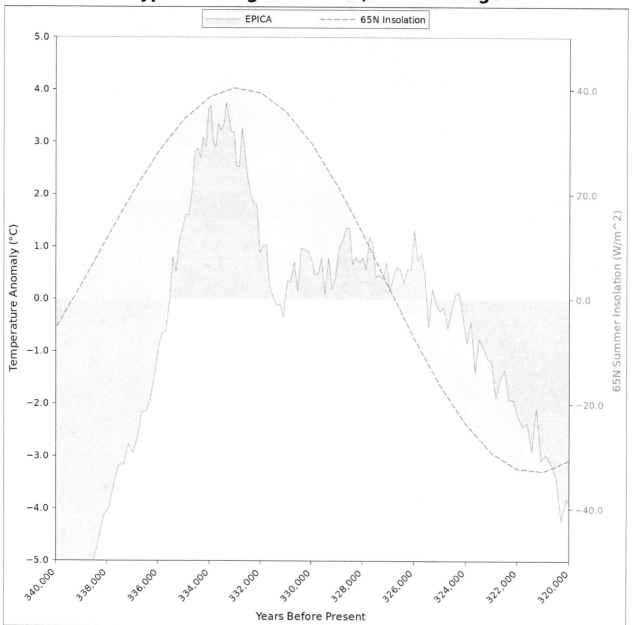

Illustration 49: Two interglacials before Eemian

This is from an interglacial that was shaped similar to the Eemian. It had a shorter duration due to less energy prior to and during the interglacial. As usual the insolation climbs prior to the warming as the ice sheets melt. The interglacial between this one and the Eemian was very short as a result of an unusually fast decrease in the summer insolation.

I used the EPICA Ice core data for this chapter because it is longest ice core and it can show the most interglacials. There are other sources of data available including the Vostok ice core and numerous sediment sets. The dates end up slightly different despite all the work to align the different sets. Since the EPICA data is the longest ice core data available that is the one I used.

Another bias issue is evident in the historical temperature anomaly charts that I have used. That is the actual temperature used to generate the temperature anomaly. What is referred to as the "normal" temperature of the Earth is no such thing. For most of the past million years the Earth has spent very little time at the current temperature or warmer. The bias is one hopelessly corrupted by the short-term view that dominates the discussion about the Earth's climate.

As I discussed in the previous chapter's scientific content, ice cores are truly wonderful proxies for reconstructing global temperature. This is primarily because most climate change takes place in the Polar Regions of the Earth and that is where ice cores are available. Well, actually Antarctica is the best place for the really old ice cores. That is because in the longest interglacial that happened 400,000 years ago melted most of the Greenland ice sheets. Even the Eemian caused some problems with the ice sheets in Greenland.

The Milankovitch Cycles:

I do apologize for using the term climate cycle instead of the proper Milankovitch Cycle. It was only done to make the introduction of the long-term cycles easier. Milutin did work in many areas from reinforced concrete (as a Civil Engineer) to the Earth's climate. He deserves the credit for cracking the cause of the glacial / interglacial cycle.

His work has come under heavy fire by many modern warmists. They consider his insolation cycle as a trigger for CO_2 to cause the warming. There are endless (truly endless) debates about the time lag from the start of the warming to the start of the CO_2 level increasing. That is because the warmist model requires that the oceans start to warm **BEFORE** CO_2 can be released into the atmosphere to cause an interglacial to form.

There are numerous attempts to try to discredit or at least weaken the Milankovitch Cycles as the cause of the glacial / interglacial cycle. Most of them center around the Total Solar Irradiance (TSI). They argue that there is little change in the TSI that the Earth receives due to the Milankovitch cycles. This ignores the most simple fact that the Earth already varies by 4 °C in the yearly cycle with nearly constant TSI. Even more damaging is the fact that TSI is highest when the Earth is closest to the Sun in January and lowest in July. So the highest TSI happens when the Earth is coldest and the lowest when it is the warmest.

That is the perfect starting point for the in-depth discussion for the details of the Milankovitch cycle.

Insolation varies as a result of changes in the Earth's orbit. There are three main factors that cause this variation[56]. They are eccentricity, obliquity and precession. I will explain the basics of each. Each one has a slightly different effect on the total insolation.

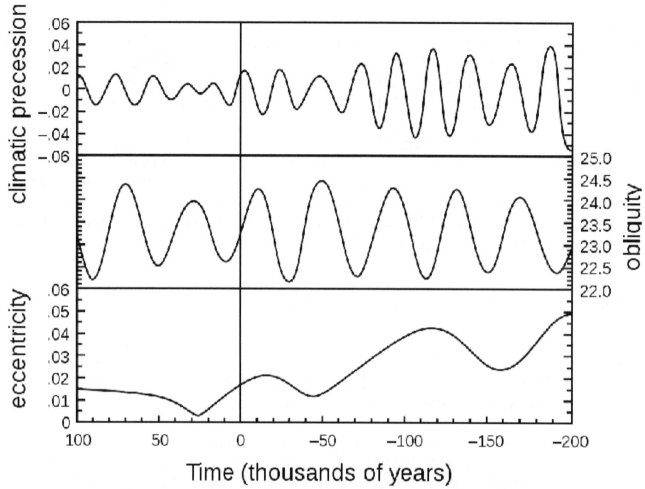

*Illustration 50: The three factors of insolation. **X-axis is reversed.***

Eccentricity seems to be the main force for climate change over the past million years. Eccentricity has 100,000 year cycles that are part of a 400,000 year cycle. The more eccentric the Earth's orbit, the bigger the distance from the sun the earth is between the closest and farthest distances. So a high eccentricity results in a bigger difference when the Earth is closest or farthest from the sun. If there is low eccentricity then the closest and farthest distances don't have as much difference in energy.

It is all about the timing. The big factor is, how close is the Earth to the sun during the NH summer? If it is closest with high eccentricity then the NH gets lots more energy. This is what happened with the Eemian interglacial. If there is low eccentricity the Earth is less likely to get high energy. The next 100,000 years will have a lower than normal eccentricity which will prevent high insolation summers from happening.

The **obliquity** is the tilt of the Earth. Over the course of 41,000 years the tilt of the Earth varies from about 21-25 degrees. A high tilt causes the NH to point more directly at the sun during the summer so it receives more energy during the summer. When the tilt is lower the NH less pointed at the sun and receives less.

The obliquity[57] was the main factor in the early part of the current ice age when the glacial cycles matched the obliquity of the Earth. That lasted for about 1.6 million years before eccentricity became the dominant forcing factor. It is not know why there was a change though. The scientists that still focus on Milankovitch (instead of CO_2 modeling) have theories as to why, but so far there has not been an answer.

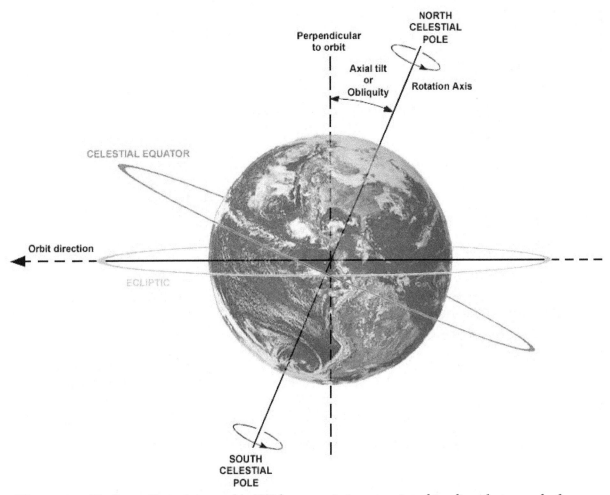

Illustration 51: Less tilt means cooler NH because it is not pointed as directly towards the sun.

Precession[58] is what determines what the season is when the Earth is closest or farthest from the sun. One complete precession cycle takes ~26,000 years to complete. So if Jan 1st was the closest approach to the sun at the start of a cycle, it would slowly change so that each day of the year (each day would average ~71 years as the closest day) until it was Jan 1st again after 26,000 years.

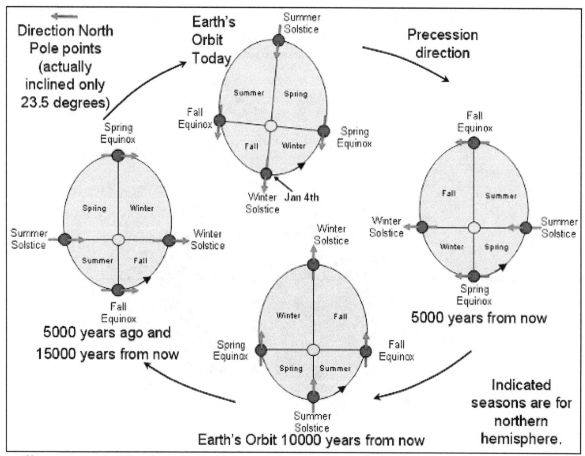

Illustration 52: How precession impacts the seasons and the insolation over a full cycle.

These three factors are not the same for each cycle as the Earth's orbit is always slightly being changed by the interaction of the Sun, Moon, Jupiter and Saturn. These more unpredictable changes add up to limit the accuracy in the past and in the future. The 5 million year data that I used is widely accepted as accurate for that range.

The Eemian was almost a perfect strike for all three factors lining up to give the maximum insolation that the NH can get. The NH was pointing to the sun while the Earth had high eccentricity with the NH summer being the closest approach to the sun. That is why it was much warmer than the Holocene.

The Holocene had good obliquity, but had lower eccentricity with good precession. The Holocene insolation peak was about 97% of the Eemian. The factors that led to the Eemian having a high insolation also gave it a very sharp drop in insolation as it went from 485 W/m^2 to

392 W/m^2. That 20% drop in insolation helped to speed the end of the Eemian faster than it would have otherwise.

It also had a smaller energy peak than the Eemian, but the energy will not drop off as far. That will probably help the Holocene last about as long as the Eemian did. Once the 65N insolation drops below 440 W/m^2 it is only a matter of time before a glacial begins. It took about 4,000 year for the Eemian to cool significantly after it dropped below 440 W/m^2.

The Holocene dropped to 440 W/m^2 3,000 years ago and has been dropping since. It took the Eemian only 1,000 year to drop as much as the Holocene has dropped in the last 3,000 years. As the Holocene is reviewed keep this in mind though. The Holocene has a flatter temperature behavior than the Eemian did. This is explained by the weaker insolation peak followed by a slower drop in insolation.

In 1976 when the paper *Variations in the Earth's Orbit: Pacemaker of the Ice Ages* was published it was considered conclusive as the cause of the glacial/interglacial cycles. Before this, there were many ardent supporters of the Arrhenius model. It would seem that many still are. This is rather confusing to me as the science on this topic is rather clear.

Every interglacial was the result of an increase in insolation. Every glacial was the result of a decrease in insolation. CO_2 always rose AFTER the temperature of the oceans rose. CO_2 always DROPPED after the temperature of the oceans dropped. This will be discussed in more depth in the chapter dedicated showing why CO_2 is a trivial and unimportant greenhouse gas.

I will speculate here a little as to why some scientists argue on behalf of Arrhenius. There are two reasons why they would want Arrhenius to be correct. First is that he is Nobel Prize winning Chemist. Arrhenius was brilliant, there is no arguing that. Milankovic was first and foremost an engineer and scientists often times do not view engineers as scientists. I expect the fact that I am engineer and not a scientist will be used against me. That doesn't really bother me because at least I am sticking to the science while many scientists are not.

There are some possible other reasons as well. Some people may have a personal bias against the use of fossil fuels and so the let their views on that skew their conclusion. There is also the money factor. Some people will always be motivated by money and there are large quantities of money involved with global warming today. This is both in the form of research grants and green-energy investment. In either case, personal bias influences their decision making process.

Chapter 7   Last Winter: The Last Glacial Period

As a skeptic I have been accused of being anti-environment. The thought being that I care more about money than I care about the world that my children will inherit. Nothing could be further from the truth. I care a great deal about the world that not only my children live in, but their children and so forth. I am a huge fan of science fiction and as such I have read many different possible futures that people think could happen. I have a very strong eye for the future of the Earth.

To quote Stephen Hawking
***"The human race shouldn't have all its eggs in one basket,
or on one planet. Let's hope we can avoid dropping the basket
until we have spread the load."***

Until such a time that humanity spreads beyond this one planet, it's all we have and that makes its health of the utmost concern for everyone. There are many very real problems facing our stewardship of the Earth, but of all the concerns out there, global warming is the absolute least concern. The reason for that has nothing to do with global warming itself. It has everything to do with the climate cycle.

The Eemian was warmer than the Earth is now and for the most part, the Earth was a better place for life. What happened after the Eemian was a far harsher world than any prediction of global warming. If you want a climate nightmare to fear, the easiest one to find is to look at what happened when the warm interglacial started to fade 120,000 years ago. It did not happen quickly as a person perceives time, but there is much to be learned from the cooling of the last interglacial.

That cooling started when the 65N insolation anomaly switched from positive to negative 120,000 years ago. When this happened the Earth was slightly warmer at that time than it is today. If the transition from positive to negative anomaly is comparable to the start of Autumn, then the Eemian Autumn started slightly more than 120,000 years ago. The change in insolation happened over thousands of years and so did the change in temperature, but over the course of 11,000 years the average temperature dropped by 8 °C[59].

If the warmist prediction is that the Earth warming up by 3 °C would be disastrous, then consider a world that was 8-10 °C cooler. That is twice the temperature difference the Earth experiences between summer and winter. That would even be a good comparison, it would be a winter, but twice as cold. It would also be winter that would not end for much of the Northern Hemisphere. That is not a prediction either, that is simply what happened when the Eemian summer ended.

Last Fall: 11,000 Years of Cooling

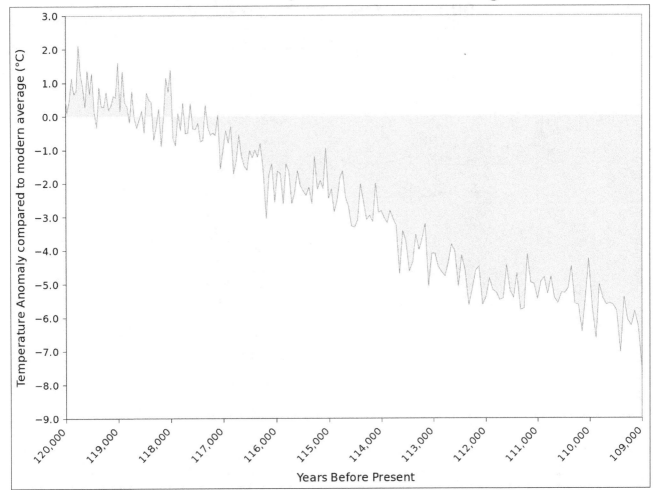

Illustration 53: For 11,000 years after the insolation anomaly went negative, the Earth cooled down. EPICA ice core.

What is interesting about this period of cooling is that it was not a period of constant cooling. There were many times when the Earth even appeared to be warming for more than 1,000 years at a time, but the long-term cooling was hidden in the short-term fluctuations. It was always present, but it was hard to see because it was small in comparison to the short-term fluctuations. There is simply no short-term period in which the cooling could be detected. The only possible way to detect the long-term trend was to look at the how the temperature was changing in periods of longer than 1,000 years.

Each 1,000 year period was colder than the one before. Most times the difference was small with an average "step" of -0.67 °C per 1,000 year period. That is a very gradual change. Such a change is not visible over the life of a single person, but spans dozens of generations. That is the rate at which climate must be observed to have meaning. If you look at each 1,000 year period, an average is just that. In every single period that is at least one time where the temperature is +/- 1 °C, and often it was both in the same period. That is the natural short-term variation in the Earth's temperature.

1,000 Year Average Temperature

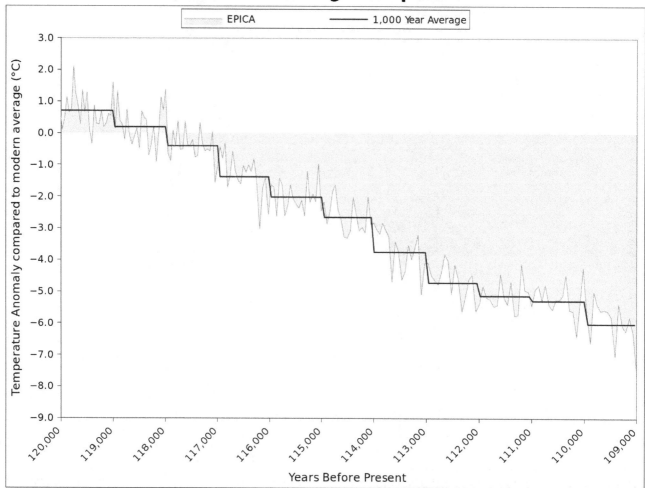

Illustration 54: Gradual is the only way to describe the long-term trends in temperature. Notice that the temperature is never actually flat. It is always warming or cooling, but the long-term trend dominates over thousands of years.

This argument will not sway an ardent warmist because they will respond to how quickly the Earth is currently warming. What this shows is that the Earth has a natural amount of variation in its temperature, even while the long-term trend is in a specific direction. In the case of the Eemian Autumn, the long-term trend was clearly downward, but the overall rate of temperature change was only 1 °C every 1,400 years. That very slow rate makes it impossible to determine the long-term trend from anything less than a few thousand years of temperature data. There is simply too much natural variation in the Earth's temperature[60].

If we refer back to the settlers in Kansas, they happened to be settling during a warm period during the Autumn. The average temperature had been increasing for a few days, but Theory #2 predicted that the shortening days would eventually result in cooling that would last for a period of months. I am going to pick a comparable period of time from the Eemian Autumn to show that

94

a couple of hundred years have the same meaning to the climate cycle as a few days does in the yearly temperature cycle.

In the 175 year period from 118,225-118,050 years ago the average temperature of the Earth rose 2.3 °C (from -0.9 to 1.4 °C). That is 3x the total warming that has been measured in the past 160 years. A large amount of warming that happened while the Earth was in a long-term cooling trend. Which trend had more meaning? Over the course of a hundred years, the short-term trend did. The rate of warming was much greater than the small long-term cooling trend. But the short-term trend lasted for only a couple hundred years. The long-term trend lasted for 11,000 years.

The natural warming that was evident 118,050 years ago certainly would have given the appearance that winter was not going to come. The Earth had been warming for almost 200 years after a period of being warm for the last 14,000 years and there was no reason to suspect that the warm period of the Earth was going to end, unless one looked at the trend in insolation. All other indicators 118,050 years ago pointed to continued warmth. CO_2 and methane levels were high. The Earth was warm and the short-term trend was to get even warmer.

The truth is that the temperature anomaly of 1.38 °C that happened 118,050 years ago would not be reached again until only 10,577 years ago. That was the Earth's last warm period for more than 107,000 years. Winter was coming and it would last for longer than our species has existed.

Of all the possible potential climate futures, the one I fear the least is a warmer interglacial. Looking back at the Eemian is all the evidence that I need to know that the Earth during a warmer interglacial is not a bad place. Sure there wasn't much good skiing because the Earth was warmer, but the Earth was a pretty nice place at the peak of the Eemian. The average temperature of the Earth was 4-5 °C warmer than it is now, but mostly that is because the Arctic was warmer than it is now. There were almost certainly thousands of years during the Eemian that the Arctic was free of ice.

So a warmer Earth is not such a bad place to live in. It is also much, much nicer than the alternative. That alternative was the glacial winter that set in after the Eemian warming had finished fading away 110,000 years ago.

Last Winter: 100,000 Years of Glacial

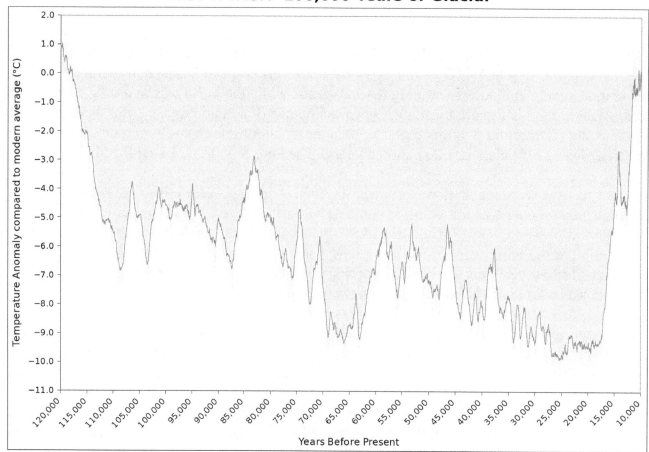

Illustration 55: For more than 100,000 years the Earth's temperature was much, much colder than the modern "normal" temperature. That "normal" anomaly is based on only 100 years of measurements. EPICA ice core.

If the predictions of global warming scare you, then perhaps you should stop reading because what happens when the Earth enters a glacial period is far worse than the most dire predictions of global warming. It isn't that one thing is worse, it is that almost everything is worse.

If the same type of cooling started to happen again, the first effect would be that the world's breadbaskets would stop producing food. The first ones to go would be the ones in Canada, Europe and Russia. The summers would simply be too cold for sustained farming. It wouldn't be all at once, but it would be yields that started to drop because the summers were just not as warm as they once were. Spring would take longer to warm up and hard frosts would happen later and later in the year.

In the fall, the frosts would come sooner and the already delayed crops would fail. The effects of this would start working their way farther south over a period of centuries. There would be times that would be slightly warmer for a hundred years here and there, but every thousand years the land that could be farmed would move further and further south.

Forget about shrinking glaciers. Let's talk about new glaciers. They would also start to form farther and farther southward from the Arctic. Mountain ranges that had been snow free in the summer would start to have more and more snow that lasted during the summer until they were covered with glaciers that would keep growing.

During the short periods of warming, they would retreat. That is how nature works. Some of the younger and newer glaciers would even disappear, for a little while at least. The long-term cooling is relentless and unyielding though. They would come back until eventually they would grow large enough that they would last for 100,000 years.

Eventually the glaciers would meet up and start to form ice sheets. At first they would be unstable and the warm periods could push them back for a while. Maybe even a few hundred years at a time. Eventually they would get big enough that they start to alter the climate of the Earth around them. No one exactly knows what thickness it takes for an ice sheet to start to alter the jet stream, but the ice sheets that covered North America grew to be kilometers thick. This had the same impact on the weather patterns that large mountain ranges have today. The ice sheet was literally a mountain range that was far larger than the Rocky Mountains are today. That same behavior exists today in Antarctica where the Gamburstev Mountain[61] range is buried kilometers under the ice sheets.

These ice sheets weigh so much that they start to push the continents deeper into the Earth. More than 10,000 years after the last ice sheets melted away the continents of the Northern Hemisphere are still rebounding from the enormous pressure that the ice sheets exerted on them[62].

Of course that much water had to come from somewhere. The only place with that much water is the oceans. During the last glacial winter the sea level dropped. Not a meter or two like we are told to worry about, but a level closer to 120m[63] (400ft) lower than they are today. Every coastal city in the world would cease to be a coastal city. In places, the coastline was more than 70 kilometers farther away than it is today.

The Earth was a very different place during the last glacial winter. Much of the modern inhabited world was under ice sheets. As far south as New York and Chicago the ice sheets covered the land. In Europe the ice sheets would cover everything from Great Britain to Poland. Most of this would be truly uninhabitable as it would be covered by crushing mountains of ice.

Dr. Blakey has shown what the Earth is like when it is in this state of existence. For most of the Northern Hemisphere this is the most difficult season of the Earth to survive. This reconstruction is from what is called the Last Glacial Maximum (LGM). It is what the Earth was like a mere 20,000 years ago. The average temperature of the Earth was 4 °C compared to the modern 14 °C.

Illustration 56: The Earth at the Glacial Maximum. 20,000 Years ago.

The above map attempts to show the physical differences for the Earth when it was at the LGM. In that regard it does a fairly good job. The little differences like Florida being twice the size that it is today because of the lower sea level is just one of the many differences. Significant physical changes also existed north of Australia and in Indonesia. The Persian Gulf did not even exist as the sea levels receded back so far.

Perhaps the most important change to the landmasses was the Bering Strait between Alaska and Asia was gone. This prevented the Pacific and Arctic Oceans from mixing. It also created a land bridge that allowed human migration to reach North America. That little fact is probably remembered from elementary school, but putting it into the context of the sea level being 120m lower than they are today is not a thought often associated with it.

The impact to the Earth's atmosphere was equally intense. There are two gases needed for most life on Earth to survive. Animals need oxygen that is provided by plants. Those plants depend on CO_2 for photosynthesis. One little discussed scientific fact about the Earth is that as the oceans cool down, they pull the CO_2 out of the atmosphere.

This is because the solubility of CO_2 increases as the water cools down. So the colder the Earth's oceans are, the less CO_2 there will be in the atmosphere. When the Eemian started to cool down, the Southern Ocean started absorbing vast quantities of CO_2 from the atmosphere that remained locked under the ice for a hundred thousand years[64]. During the last winter's glacial period the CO_2 levels dropped as low as 180 ppm[65].

This caused many of the Earth's forests to die off from lack of CO_2. The greatest damage to ever happen to the rainforests was caused by CO_2 levels that were too low for them to breathe[66].

This caused most of the Earth's rainforests to be replaced by grasslands. Grasses can survive with lower CO_2 levels and so they survived while most of rainforests almost completely disappeared around the world. There is nothing more catastrophically destructive to rainforests than not having enough CO_2. When people worry about deforestation of rainforests, there is nothing that will do it better than an advancing glacial that cools the oceans enough to pull the CO_2 out of the atmosphere.

The rainforests that existed in Brazil and Africa during the last glacial period were reduced to tiny little ones that were surrounded by vast grasslands that separated the rainforests from each other. So well-known is this amongst scientists that one of the leading theories for the biodiversity in rainforests is thought to be caused by species of animals evolving in the isolated rainforests during glacial periods.

This leads to another fact. The Amazon rainforest as we know it is young. It did not exist for most of the past 110,000 years. For most of that period the region that is now the Amazon rainforest was mostly a savanna or prairie. The Amazon was likely much the way it is today during the Eemian, but when the Eemian ended, so did the rainforest that existed there. The rainforests in Africa also shrunk in a similar manner[67] and the deserts that exist now were much larger during the last glacial period.

Australia was also drier and large portions were covered by rolling sand dunes[68] that dominated the interior. Much of the Earth was without plants during large portions of the last glacial period. It is difficult to separate out what was caused by lack of moisture and what was caused by lack of CO_2 and there is debate about which effect was stronger, but it is thought that the tropical oceans were of comparable temperature to what they are today. So there was still plenty of moisture in the atmosphere to cause precipitation to exist in the tropical regions of the world. There is tantalizing evidence[69] that the tropics are even wetter during the glacial periods. There is still much to learn though.

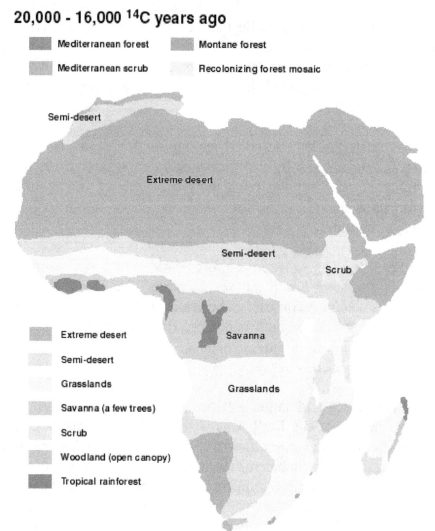

20,000 - 16,000 ¹⁴C years ago

Mediterranean forest   Montane forest

Mediterranean scrub   Recolonizing forest mosaic

Semi-desert

Extreme desert

Semi-desert

Scrub

Extreme desert

Semi-desert

Grasslands

Savanna (a few trees)

Savanna

Scrub

Grasslands

Woodland (open canopy)

Tropical rainforest

*Illustration 57: During the glacial periods the rainforests shrunk
to a fraction of what they are today.*

Also consider the fate of the coral reefs that lived during the Eemian. Every single tropical coral reef that was alive during the Eemian died when the glacial started to form. Not a single coral reef survived the arrival of the glacial. All the life that lived in the reefs also died. Imagine a period so brutal that all the coral reefs on Earth died in the period of a few thousand years. Not one or two, but all of them.

This is a known fact because the sea levels dropped so dramatically at the end of the Eemian. Since coral reefs can only live at a certain depth of water and temperature, the knowledge that the sea levels dropped more than the depth that coral reefs can live in is sufficient evidence that all the coral reefs from the Eemian died when the last interglacial ended.

Of course the coral reefs re-established themselves at the lower sea levels, but the sea level was not all that stable during the course of the glacial period. Interestingly enough the Earth's oceans kept changing their level in accordance to the amount of insolation that 65N was getting.

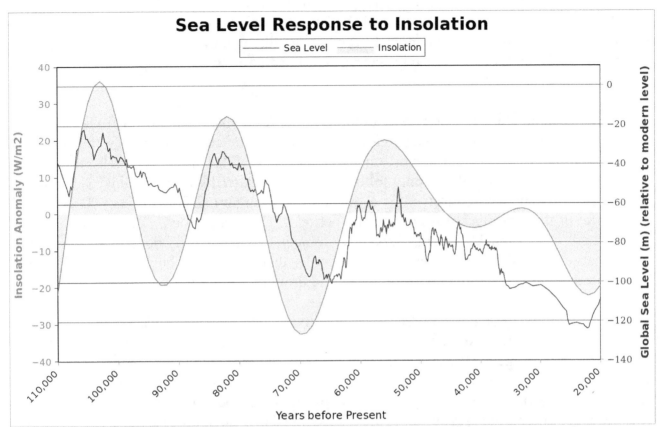

Illustration 58: Global Sea levels changed constantly with the changing insolation anomaly at 65N. In periods where the energy increased, the glacial would retreat slightly. When the insolation dropped, the glacial advanced. All of this is reflected in the sea levels.

70

Every time that the sea levels changed by 10-20m, most of the world's coral reefs would have died again. So not only did the coral reefs die once when the Eemian ended, there have probably been 6 or 7 coral killing changes in sea level during the last glacial period. Coral reefs would have died each time the sea level dropped enough to expose them to the atmosphere. In addition, whenever the sea levels rose by 20m it would have blocked enough sunlight to kill most tropical coral reefs. The only conclusion is that the last glacial period wiped out all of the Earth's coral reefs many times in the past 120,000 years. Fortunately coral reefs have adapted to the constant change of the Earth's sea levels and are always expanding to places where they can survive.

Two of the most commonly discussed ecosystems supposedly at risk for global warming, the coral reefs and rainforests, suffered the most when the Earth cooled down after the last interglacial. It is also worth noting that both the rainforests and the coral reefs did very well during the Eemian when the Earth was warmer. From the perspective of the coral reefs and the rainforests it would be best if the Earth stayed exactly the same as it was during the Eemian. Unfortunately for life during the Eemian, the Earth did not stay the same. It got much, much colder.

There is one period during the last glacial period that was even more difficult than the rest of it. It was the period between 25,000 and 45,000 years ago. This was a period of unusual insolation from the sun. The 65N insolation anomaly can behave like this every 400,000 years or so. The last time it happened there was a long interglacial, but this time the energy was lower and the result was disastrous.

Instead of stable climate it brought about one of the least stable climates that we have yet to discover. The ice cores from Antarctica do not show as much variability as the ones from Greenland do, but even they show more change than usual. What the ice cores from Greenland show is short periods of rapid warming followed soon thereafter by rapid cooling. According to the ice cores the temperature changed by as much as 4 °C in periods as short as 100 years. The worst aspect of these changes was how short a period of time they lasted. The NH would warm, and then cool just as quickly. These spikes in temperature happened about 10 times over that 20,000 year period. The Earth remained much cooler than it is today, but such unstable climate likely caused rapidly changing periods of floods and droughts.

It was during this period of rapidly changing climate that mankind's closest relative (possible ancestor for most people) went extinct. The more that is learned about the Neanderthal the more intriguing their story gets. What is known is that this durable branch of modern man survived many glacial / interglacial cycles while living mostly in what is now Europe. Despite this they did not survive this period of extremely unstable climate. It has been proposed[71] that this unstable climate was the cause of their demise. I am a bit skeptical of that claim, but it is certainly possible that the rapidly changing climate did play a role.

This leads to another claim about global warming: that climate is changing at an unusually high rate[72]. That is an absolutely absurd statement. In the past 150 years the average temperature has changed less than 1 °C. During the period of unstable climate during the last glacial period it could change 6 °C in less than 150 years. That size of change occurred in both flavors, warming and cooling, although the cooling was slightly slower than the warming.

What is not found in the 100,000 year glacial is a period where the climate stayed the same for long periods of time. That has never happened, not even once. There might be times when it happened to be the same temperature at periods 1,000 years apart, but the Earth's temperature did not stay constant for any significant period of time.

As far as I have found, there has never been such a period in any of the temperature reconstructions. It would appear that the Earth is never really staying at the same temperature. There are periods where the scale of the temperature change is less than at other times, but that temperature window of variation has ALWAYS been more than +/- 1 °C over the course of a few hundred years and certainly for periods that are longer than 1,000 years.

This idea that the Earth's temperature is stable and that there is a correct temperature is the greatest single misconception that exists about the Earth's climate. The Earth has spent less than 10% of the past 100,000 years warmer than it is now. Over the past million years it has been this warm for much less than 10% of the time. To call the average temperature of a 100 year period the correct temperature would be EXACTLY like measuring the average temperature of the Earth from noon to 8:45pm on October 15th and calling that the only correct temperature.

The Earth's temperature was not stable during the Eemian, nor was it stable during the last glacial period. After the period of unstable climate ended 25,000 years ago, the Earth got even colder still. It cooled down to what is called the Last Glacial Maximum (LGM) That was the greatest extent of the ice sheets and the lowest level of the oceans.

At the LGM the sea levels were so much lower that an additional 26 million km^2 (10 million square miles) of land was above sea level. That is comparable to adding another continent the size of North America to the Earth's landmasses. Conversely, this means that in the past 20,000 years the oceans have risen up and covered a continent worth of land.

So even during a glacial period that lasted 100,000 years there was near constant change that was either warming or cooling. The oceans rose and the oceans fell. Rainforests and coral reefs suffered and died. Deserts expanded and retreated. Glaciers and ice sheets also advanced and retreated. The composition of the atmosphere varied as well.

Extinctions were also happening throughout. Mankind will never know how many species of life died during the mass deaths of the coral reefs at the end of the Eemian or the collapse of the rainforest ecosystems that happened at the same time. We do know that the Neanderthal, who was one of the most intelligent species to live on Earth failed to survive the last glacial.

All of this climate change was caused by the variations in the energy the Earth was getting from the sun, especially at that zone of 65N during the summer months. In the current geographical configuration of the Earth, that is the zone that drives real climate change. The variation in the insolation at that one zone has the power to change the Earth's sea levels and with that power it can alter the climate of the entire Earth.

All of this has happened before. All of this will happen again.

Warning:

Scientific Content!!!

It is not possible to give the last glacial period adequate coverage in a single chapter or even a single book. So I stuck to the highlights and topics that are in the global warming debate. Many of the dire predictions about global warming are simply based on events that have already happened many times in the Earth's past. The problem is that the warmer climate is always better for life on Earth than the colder climate.

The start of the glacial period (often referred to as the Wisconsian Glacial) is useful in many ways. It shows that elevated CO_2 levels don't prevent the Earth from cooling down, but even more useful is that period had a steady long-term rate of temperature change. That allows the temperature to be de-trended so the statistical behavior can be determined.

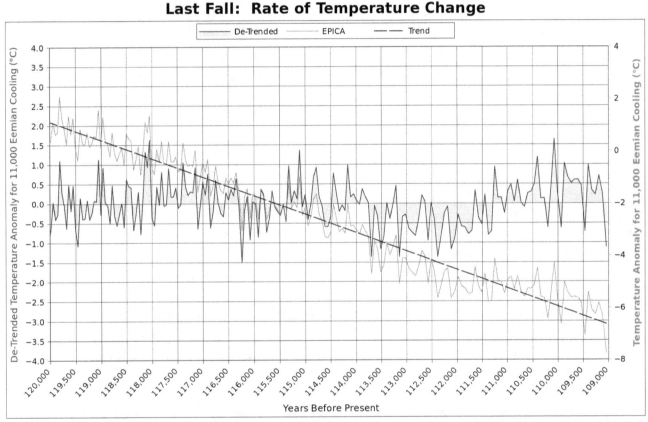

Illustration 59: (Blue) EPICA temperature anomaly is subtracted from the (Black) trend line which results in the (Brown) de-trended temperature of the Eemian cooling.

The resulting data has no linear trend so it can be used to estimate what the natural variability is of the Earth's temperature. This is critical because it allows us to know how much the temperature of the Earth can be expected to change in the short-term while moving in the long-term trend.

The result is a standard deviation of 0.59 °C. So in technical terms anything that is within +/- 1.18 °C is within the statistical noise. The precise definition is that 95% of the data will fall into that range. Visibly it is much easier to see what is going on.

Last Fall: Natural Short-Term Variation

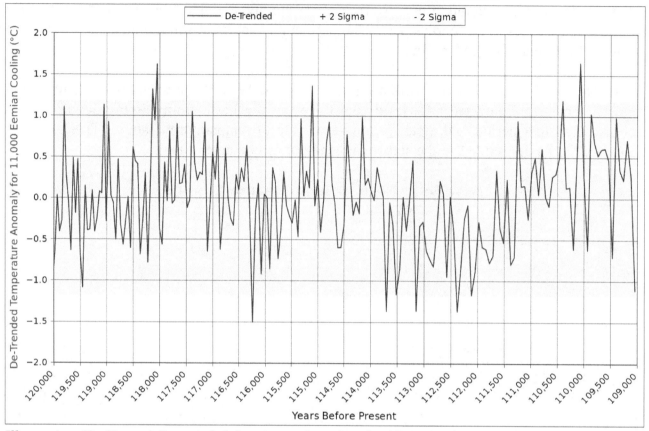

Illustration 60: (Brown) De-trended data is mostly within the (Green) +/- 2 Sigma. This shows that the natural variation of the Earth's temperature is at the minimum +/- 1.18 °C.

Any trend within the green region is basically meaningless noise. While that noise has a real world impact, it is part of the natural variation of the Earth's temperature. It is also worth noting that this is the variation of a 55-year average. That means that in periods of shorter than 55 years there would be even more variation in the temperature.

The 55 year average by itself is statistically devastating to the theory of global warming. When it takes a 55 year average to exceed the +/- 1.18 °C 2-sigma natural variation there is no statistical basis to state that the Earth has experienced any statistically significant warming. The total variation in the past 160 years is +/- 0.4 °C which is within a single standard deviation. Even that overstates the case of warming because that is using single year data and comparing it to 55 year averaged data.

That the short-term trend has been warming is scientifically interesting, but when compared to natural behavior of the Eemian interglacial it is clear that the recent warming is insignificant in magnitude and duration. If the warming is within the natural level of the noise, then no valid argument can be made that it is anthropogenic in nature.

What scientists should be asking is why does the Earth naturally vary by 1.18 °C over the course of centuries? That has not happened. Instead of looking at why the temperature naturally varies there has been more and more research into trying to explain how CO_2 levels have caused what appears to be purely natural variation in the Earth's temperature. The most fundamental problem is that scientists have stopped asking the correct questions.

CO_2 Solubility in Water

If the solubility of CO_2 was different than it is, there would be no global warming debate. There would also be no life on Earth as we know it today, so it's a bit of a double-edged sword. The facts are very simple to this. Water that is cooling takes CO_2 from the atmosphere and water that is warming puts CO_2 into the atmosphere. That is simple. No one can scientifically argue that most basic fact.

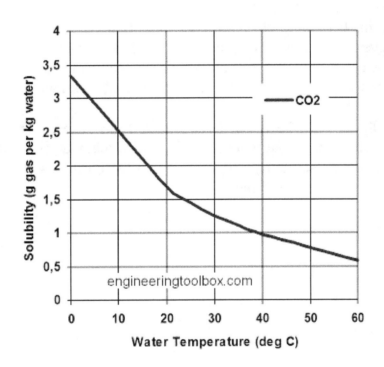

Illustration 61: Colder water absorbs more CO_2 than warm water. Near freezing water absorbs __twice__ as much CO_2 as 20 °C water does.

The confusion about temperature and CO_2 arise from the fact that the amount of CO_2 in the atmosphere depends on the temperature of the ocean. This is why the level of CO_2 in the atmosphere changes with temperature. CO_2 does not cause temperature change, it simply changes based on the temperature of the ocean. This is why the level of CO_2 changes AFTER the temperature has changed. It does that for both warming and cooling. A cooling Earth removes CO_2 from the atmosphere so the CO_2 level drops. A warming Earth does the opposite and releases

CO_2 into the atmosphere which increases the CO_2 level.

This of course results in the level of CO_2 in the atmosphere fluctuating based on the temperature of the oceans. It also means that the CO_2 level rises and drops AFTER the temperature does. This very simple and basic fact is precisely what is observed in nature. Every ice core shows the same exact behavior. CO_2 levels change AFTER temperature change happens.

The real problem is that a theory was proposed before anyone was able to measure global CO_2 levels and link that to the Earth's temperature with any accuracy. Once they looked, they found that CO_2 levels did in fact fluctuate with the Earth's temperature. Since the theory proposed by Arrhenius predicted this, it was considered proof that he was right. Unfortunately there was no way that this wouldn't be the case since the solubility for CO_2 in water will ensure that it will always be true.

This is the greatest mix-up of cause and effect to ever happen in science. Generally the debate rages about the lag at the beginning of the interglacials. Everyone else ignores the fact that CO_2 levels are always elevated when the interglacials end. If CO_2 caused the warming in the first place, it would also be able to prevent the cooling. In all the history of the Earth there has never been an instance where CO_2 has prevented the Earth from cooling down.

In the past 720,000 years the Earth has experienced 8 glacial periods. Each of them started when CO_2 levels were elevated. What more proof is needed that CO_2 is not the thermostat for the Earth? This single section on solubility and the fact that the Earth has always cooled when CO_2 levels are high should be enough to end the debate.

Chapter 8 Holocene Interglacial

During the last glacial period humanity was able to expand throughout the world. Since the continental ice sheets covered much of the Northern Hemisphere, it was the places that were not covered with ice that were available. The lower sea levels and expanded land mass made the migration of humanity to most corners of the Earth possible. It was the Holocene that allowed civilization to flourish. It was the perfect opportunity for the highly adaptable humans.

The transition from the Last Glacial Maximum (LGM) to the Holocene was dramatic. There is little known about the people that lived in that time because it has been lost. Much like people today live by the oceans it is likely that many of the people during the glacial period lived by the oceans. It is the ocean's story of an interglacial that really explains how much the Earth changes. Rising temperatures are interesting, but an ocean that rises by at least 20m (66ft) in a period of as little as 200 years is real climate change. That massive change in sea level[73] is known as the first major Meltwater Pulse 1A (MWP-1A) that took place ~14,500 years ago[74]. The continental ice sheets were responding to the increasing insolation from the Sun and they were melting.

Such a change would have resulted in a visible yearly change in the ocean's level. In a period of 10 years the oceans would have risen as much as a meter. This was an increase that would have continued relentlessly for many generations. Living by the ocean would have been a series of moves every few years to stay ahead of the ever rising ocean.

Any cities[75] that existed at the beginning of the MWP-1A would have been buried under the ocean within a single lifetime. The most commonly accepted time for MWP-1A was ~14,600 years ago. The 65N insolation at that point was rapidly increasing and was already above normal. The insolation kept increasing and so did the sea level.

There were 3 major melt water pulses as the ice sheets and glaciers melted because of the increasing 65N summer insolation. In total the sea levels increased more than 100m (330ft) over a 10,000 year period. That change in the sea level was not a steady change. Most of that change happened rapidly during the 3 melt water pulses.

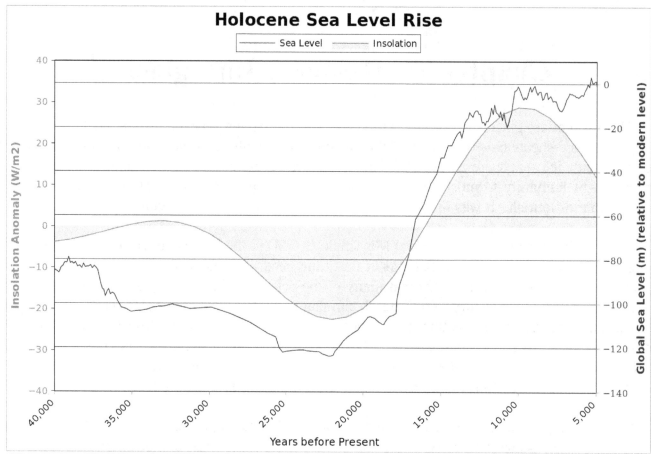

Illustration 62: When the 65N insolation started increasing 20,000 years ago the continental ice sheets melted. In a few thousand years the sea level increased nearly 100m (330 ft).

If the sea level was changing at the same rate today as they did during the largest of the pulses, the sea levels would be ~7m (22ft) higher today than they were in World War II. The actual change in sea level since WWI is ~0.1m[76]. The oceans were rising 70 times faster than they are today. Such a change in sea level today would be catastrophic. I will be the first to admit that. The problem isn't that the sea levels are changing though. The problem is our perception that the sea levels should stay the same. The reality is that the sea levels have always been changing.

Today the rate of sea level change is so small that it takes satellites to accurately detect it. There are several places that provide the data for anyone to review[77]. During the early Holocene the sea level was rising so quickly that in a single lifetime the ocean would engulf most coastal communities on the Earth today.

That is climate change. That is also part of a very natural cycle that has been repeating for hundreds of thousands of years on the Earth. When I hear a dire projection about the sea levels changing, I have to laugh. The sea levels have never been constant. They have always been changing. Much like rivers and lakes rise and fall over the seasons and years, the sea levels are always changing. They do not change as quickly as the rivers do, but the oceans themselves are still always changing.

110

It is no surprise that all cultures around the world have ancient myths of great floods[78] in their past. While the modern world is fretting about a "possible" change in sea level of a meter or so, our ancestors survived oceans rising by meters in a person's lifetime. This doesn't include the floods of water traveling from the ice sheets to the oceans. The maps of the world were re-drawn not once, but many times during that period of transition from glacial to interglacial.

All of that change was natural. In no way could humanity have caused it, but it did survive it. We will never know the cost of that survival. All we know is that mankind was mostly a tribal hunter-gatherer society at the end of the glacial period and when the Holocene finally stabilized some 7,000 - 9,000 years ago, modern civilization was born. Much of the evidence of earlier civilizations that existed before that is now buried hundreds of feet under the oceans. Coastal regions that are as large as the North American continent are now underwater. It is likely that people lived on most of that land as recently as 18,000 years ago.

It wasn't just the sea levels that were changing, the Earth was also rapidly warming up. More than any single thing it was the positioning and conditions of the Early Holocene that allowed humanity to transform itself from a hunter-gatherer society into an agricultural one. The conditions of the early Holocene were warmer and more moist than they are now[79]. As a result, farming was easier than it is now. The early phases are called the Holocene Climate Optimum.

When it comes to dealing with the ever changing climate of the Earth, mankind has some real experience. The transition from the LGM to the Holocene Interglacial was about as severe a change as the Earth experiences in periods of 5-10 thousand years. The average temperature of the Earth increased by as much as 10 °C with much of the temperature change taking place in the Polar Regions. In the northern mid-latitudes the climate also changed greatly as the ice sheets from the last glacial melted. Even the amount of ice in the Southern Ocean surrounding Antarctica saw significant decrease in the long-term sea ice that melted as the Holocene warmed up.

Holocene Warming: Current Interglacial Warming

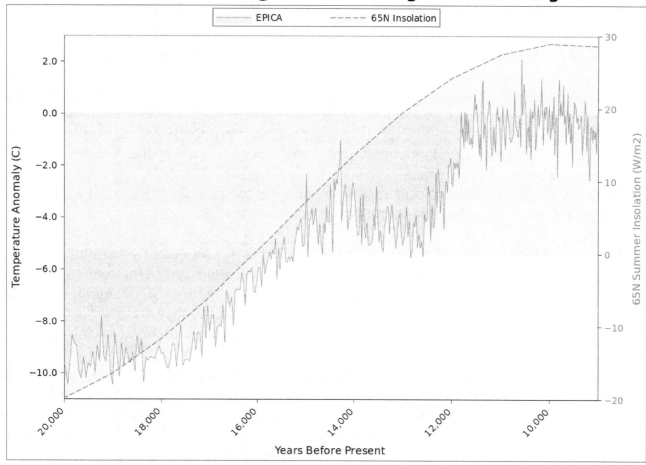

Illustration 63: The Earth's temperature responded to the higher summer insolation in the Northern Hemisphere by warming. This caused the glacial continental ice sheets to melt early in the Holocene Interglacial.

Even this dramatic temperature change saw a very low long-term warming rate of only 10 °C over a period of 6,000 years. That is only 0.167 °C per century or 600 years for a 1 °C change in temperature. Many people will read this and argue that the current rate of warming is higher than that, but that would not be a valid comparison. The long-term rate of temperature change is always small and that is why it is hard to detect it. There were many centuries with very high rates of warming and even centuries with high rates of cooling.

Earlier I showed the climate cycle comparison for the Eemian and the Holocene Interglacials. It is time to take a closer look at that chart.

Interglacial Comparison

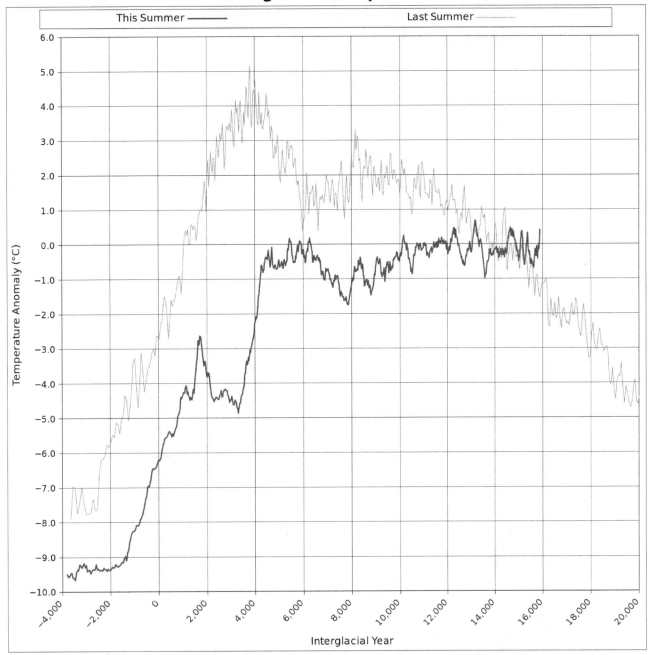

Illustration 64: The Holocene and Eemian interglacials. EPICA (Antarctica) ice core.

There is an obvious difference between the two interglacials. The Eemian warmed up more quickly of the two. In addition it warmed up at least 4 °C more than the Holocene did, at least according to this ice core. The Holocene has been the more stable of the two. At the same point the Eemian was already cooling down while the Holocene is not showing the same rate of cooling that the Eemian did at the 16,000 year mark. These differences are substantial, but fortunately the Milankovitch Cycle provides a good explanation.

The answer not surprisingly is insolation. The complexity of the Earth's orbit causes the insolation behavior for each cycle to be different. If the insolation behavior for the summer insolation at 65N is shown in the same manner as the temperature anomaly the cause of the temperature difference is easily shown.

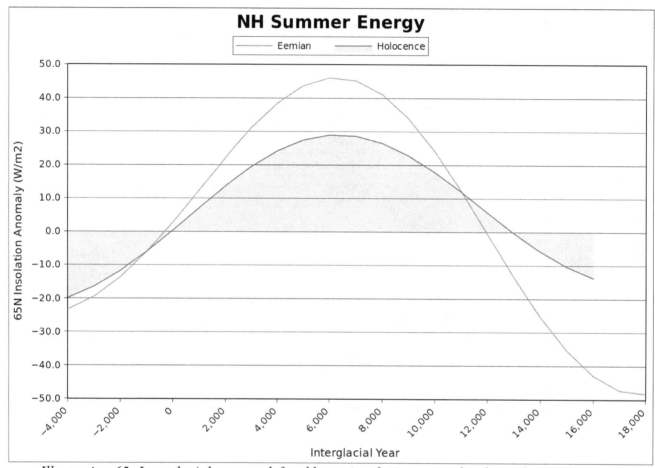

Illustration 65: Interglacial year as defined by an insolation anomaly of zero for 65N summer insolation. The Holocene is currently at interglacial year 16,000.

The Eemian experienced a faster increase in the insolation and the peak insolation was ~50% higher than the Holocene. That the Eemian experienced a 50% more warming than the Holocene. In addition the Eemian insolation dropped off more quickly which would explain why the Eemian cooled more quickly. In each difference in temperature, the insolation provides a scientific cause for the different interglacial behavior.

The Holocene will have positive a insolation anomaly for ~1,000 years longer than the Eemian did. The rate at which the insolation is dropping is also lower at this point of the Holocene. The result of that difference is that the Earth will not cool as quickly as it did during the Eemian. That is very good news because it the Eemian cooled down relatively quickly. The bad news is that the insolation anomaly went negative 3,000 YBP. The only possible result of that is cooling.

114

Unfortunately there are not many ice cores in the NH that allow a good comparison between the interglacials. So I had to use the Antarctic ice cores for the interglacial comparison. The EPICA one in particular is good for that purpose, but it is also the least sensitive to changes in the NH. To show what is happening in the NH I need to show the Holocene from the perspective of the NH. To do that I will use the GRIP (Greenland Ice Core Project).

Holocene Interglacial: Greenland

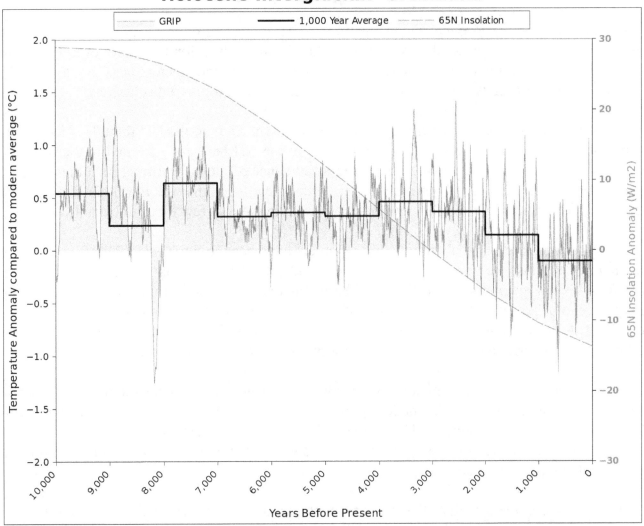

Illustration 66: GRIP (Greenland Ice Core Project) reconstruction of the past 10,000 years.

Once again there is significant short-term variation, even more than in the Antarctic ice cores, but there is also a long-term trend in Greenland[80]. 3,000 years ago the insolation anomaly for 65N went negative and each 1,000 year period since then has been cooler than the one before. I also notice that the last 1,000 year period is the coldest 1,000 year period in the last 10,000 years. It could be coincidence that the 1,000 year period that has been getting the least 65N insolation also happens to be the coldest 1,000 year period of the past 10,000.

It could also be coincidence that the 2nd lowest insolation period
is the 2nd coldest 1,000 year period.

115

It could even be coincidence that the 3rd lowest insolation period
is the 3rd coldest 1,000 year period.

But it does not appear to be coincidence. It appears that since the insolation anomaly went negative 3,000 years ago that the Northern Hemisphere has started to cool down. If I look at the long-term trend then the **very last** thing I am worried about is global warming.

It is not just one ice core that shows this in the NH, it is all of them. Any glacier that existed during the last glacial period shows that there has been cooling for the past several thousand years. That is far too important to be ignored. If the data from the glaciers is saying that the NH has been cooling for thousands of years, why isn't that the main topic of discussion?

There is an even more important lesson being told by glaciers in the NH though. That story is even more important to understand. Most glaciers in the NH have only formed in the last 4,000 years. While these new glaciers are often used as evidence of global warming, that these glaciers have formed in the past few thousand years is actually evidence that the Earth and the Northern Hemisphere specifically are cooler now than they were in the past few thousand years.

If most people are asked about how long glaciers have been around, they believe that they are remnants of the last ice age. Very few people understand that this is not the case. It is not something that is given much attention, but the fact is that when the last glacial period ended and the Holocene began, very little ice outside of the Antarctica and Greenland survived the warmer periods of the early Holocene. Most of the little glaciers that dot the NH today are new glaciers that have formed since the insolation anomaly went negative 3,000 years ago.

A glacier that is only a few hundred years old will always be at risk for the normal variability of the Earth's temperature and the modern period is no exception. Some of these new glaciers are already hundreds of meters thick. That is the real story of the current state of the Earth. The NH has been growing glaciers for thousands of years now. That only happens when the warmth of an interglacial is starting to fade.

In many cases the scientists that have been studying glaciers expected to find that they did exist from the previous glacial, but ended up finding that the glaciers had only recently formed. That is exactly the type of unexpected discovery that should cause scientists to re-examine the theory of global warming, but so far that has not yet happened. The story of the glaciers is telling us that the Earth was warmer a few thousand years ago and is now cooling down. The rate is not fast, but it is steady. This is exactly the type of behavior that the Eemian interglacial experienced as it was ending.

So I am going to provide information about 4 glaciers that have formed in the past few thousand years. The older glaciers formed farther north and the youngest one only recently formed more than halfway to the equator. The implication that glaciers have been forming farther and farther south as the insolation has been decreasing is the greatest indication of what is really happening to the Earth's climate for the long-term trend. The glaciers that I will discuss are located in Northern Greenland, Alaska, Southern Canada and finally in the state of Wyoming in the United States. These glaciers are always in a state of change, sometimes growing and sometimes shrinking, but none of them are remnants from the last glacial, which means that they have all formed as a result of the cooling climate of the Holocene.

The first new glacier I will cover is in Northern Greenland, but it is not part of the main ice sheet of Greenland. It is known as the Hans Tausen Ice Cap in northern Greenland. It has been drilled several times and bedrock was found at a depth of 325m. This glacier is ~4,000 years old[81] and no part of it has ever been found to be older than that. This glacier started to form even before the insolation anomaly went negative in the NH.

Illustration 67: The Hans Tausen Ice Cap in Peary Land, Greenland is 4,000 years old. It is separate from the main ice sheets that cover Greenland.

This one is very important because it also clearly shows that this region of Greenland was warmer 4,000 years ago than it is today. This shows that the main Greenland ice sheet survived thousands of years of warmer climate than it is experiencing now. If a glacier has steadily been growing in Greenland for 4,000 years, it would indicate that the main ice sheet has also been growing over the same period of time. The GRIP ice core also shows that the Earth was warmer for most of the past 10,000 years than it is today. If it survived 10,000 years of warmer climate already, there is little reason for us to worry about it today.

In 2002 Lonnie Thompson expected to find an ice core that was at least 18,000 years old at Mount Churchill (a few hundred kilometers East of Anchorage, Alaska). He has done ice cores around the world based out of Ohio State University. What he found surprised him and many others. The oldest ice at the bottom of the 460m (1500ft) glacier was only 2,500 years old[82]. Meaning that mountain was ice free prior to that. That region of Alaska has cooled enough in the past 2,500 years to grow a glacier that is 5 football fields thick and 64 km long (40miles).

Illustration 68: The nearly half kilometer thick glacier between Mount Bona and Churchill is ~2,500 years old. Strong proof that this area has cooled in that period of time.

This glacier is farther south than the one in Greenland and it is also 1,500 years younger than the one in Greenland, but it has been growing faster and is already thicker than the Hans Tausen Ice Cap. The primary cause of this would be more precipitation in Alaska than in northern Greenland.

Moving even farther south into the lower part of Canada is the Mt. Logan Glacier[83]. This is the 2nd highest mountain in North America. It would even be reasonable to expect that ice from the last glacial period survived there, but the fact is that the oldest ice ever found was about ~8,000 years old. This is older than what was found in the previous glacier in Alaska, but in this case altitude is a major contributor to the climate of Mt. Logan. The upper plateau is over 5,000m in altitude and as such is very cold year around. Yet even this mountain lost it's ice in the main warmth of the Holocene. The ice has been growing there and is currently 300m deep in spots. Most of the ice here is less than 300 years old according to the main ice core that was drilled in 1980[84].

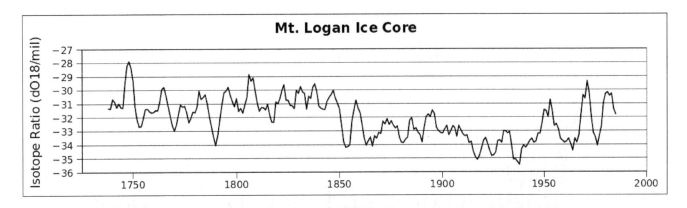

Illustration 69: Mt. Logan in Canada was also drilled to bedrock in 1988. Many parts of the glacier are only a few hundred years old, but in places it is up to 8,000 years old.

The youngest of these new glaciers is the Fremont Glacier[85] in Wyoming. This one is a very young glacier and the ice from it is generally only a few hundred years old at most. It likely started to form during what is called the little ice age that took place a few hundred years ago. Even more significant is the location of this glacier. It is at 43°N latitude. This is an indication that the Holocene has cooled enough for glaciers to start forming more than halfway to the equator.

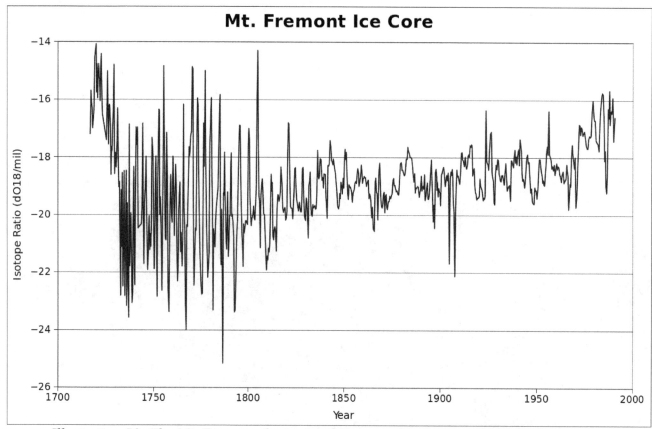

Illustration 70: The Mt. Fremont glacier is ~300 years old. It is nearly 160m thick in that short period of time.

This southward trend of glacier formation is evidence far more powerful that proxy temperature reconstructions could be. The conditions needed for glaciers to start forming are simply summers that are cold enough to prevent the snow from melting. 1,000 years ago it was too warm in Wyoming for this particular glacier to form. In the past 1,000 years it became cold enough for this glacier to form and many others like it around the NH.

It isn't only in the NH that glaciers have been forming in the past few thousand years. One of the more famous glaciers in the Southern Hemisphere is the Quelccaya Ice Cap in Peru. That puts this series of glaciers in the tropics. While the glaciers here have been well documented to be retreating in recent years, it is far more interesting that the ice cores from it only date back to 470 AD, which makes this glacier complex in the SH a mere 1,500 years old[86]. This shows that prior to that, it was warm enough that no glaciers were present here.

This particular glacier complex is in the tropics and is the largest glacier system in the tropics. Altitude plays a large role in the ability for glaciers to form in the Tropics at any time and this is no exception as the average altitude here is over 5,000m. That does not take away that this ice cap did not exist 2,000 years ago, but it exists now. At only 13°S latitude the climate has cooled enough to allow the formation of large glaciers in the last 2,000 years.

Illustration 71: The Quelccaya Ice Cap in Peru has been used repeatedly as proof of global warming as it has retreated in recent years. What isn't mentioned is that the ice cores to bedrock only date back 1,500 years.

These long-term changes are subtle and take place over hundreds of years, but there is no greater evidence that the Earth has been cooling for thousands of years already than the formation of glaciers where none existed before. The formation of these glaciers is increasingly closer to the equator as well. That is the signature that autumn has arrived for an interglacial.

The younger the glacier is the more recent the cooling happened that allowed the glacier to form. There are other glaciers throughout the world that are only a few hundred years old. These also happen to be the glaciers that are currently showing the most impact from the current warming variation that is happening.

This is no different than an early snow that melts away after a few days in the yearly cycle of the seasons. That snow melts in October has never been proof that winter would not arrive. On the scale of the climate cycle there is absolutely no difference in a glacier that forms and melts away on the hundred year scale. The only problem is mankind's ability to perceive it for what it is.

In every way the Earth is acting exactly as it should for this stage of an interglacial. Much of the Earth is still warm, but the higher altitudes and latitudes are starting to see the effects of the long-term cooling. It is not a rapid onset of a glacial like many people would expect. In every way the climate cycle acts like a longer term version of the yearly cycle.

The year's first snow is at the high altitudes and far to the north. As the winter progresses the snow level gets lower and lower until the valleys and lower altitudes are covered with snow. Glaciers follow the same pattern of formation. They started to form in the far north and at high altitudes. The older glaciers are now hundreds of meters thick. They have continued to form farther and farther south. Over the past 1,000 years they have started to form halfway to the equator. These glaciers grow at times and shrink at times; much like the snow level will rise and fall in the mountains during the Autumn each year. Over thousands of years the net effect is that they get larger and larger until eventually they are large enough to alter the Earth's climate and a full glacial period arrives.

The glaciers tell us the story of the Earth's climate. Part of that story is that the last 1,000 years have been the coldest 1,000 year period in the last 10,000 years.

Let me repeat that.

The last 1,000 years have been the COLDEST of the past 10,000 years.

Then I am told that warming is the greatest threat facing the Earth today because in the past 150 years the average temperature of the Earth has varied by the amount that is perfectly normal. There is no evidence in the climate data that anything unexpected is happening. The only thing that is evident from the data is that the Holocene Interglacial is starting to fade. Glaciers exist now that did not exist 1,000 years ago.

The only promise that can be made by anyone is that the Earth will not stay the same as it is today. Never in its history has the climate been constant. The Earth is always warming or cooling. That is natural. When I compare the autumn of the Eemian to the Earth's climate right now, they are as comparable as any two years are to each other. They are not identical, but the tendencies are the same. There is not an immediate drop in temperature; there is a gradual drop where the temperature drops close to a degree each 1,000 year period. That is exactly what the Earth is doing today.

All of this has happened before. All of this will happen again.

That is the nature of these cycles. The only difference now is that mankind has a chance to understand the cycle before it is too late.

Warning:

Lots of Scientific Content!!!

Any objective and neutral assessment of the Holocene Interglacial will reach the conclusion that the peak warmth occurred 6,000-8,000 years ago. Various places give slightly different results, but overall the peak warmth of the Holocene has long since past. Today the Earth is cooler and the glaciers are closer to the equator than they were only 1,000 years ago. There is also very little room for disagreement that the climate started to cool ~4,000 years ago. Since that point 4,000 years ago, every 1,000 year period has been colder than the one before. That is precisely the behavior that the Eemian showed after the insolation anomaly went negative.

Just like each November is colder than October, the interglacial fall is cooling in the same manner. The last 1,000 year period has been a cold one. Glaciers have started to form where there have not been glaciers for nearly 10,000 years. That is not evidence of a long-term warming trend; it is absolute evidence of long-term cooling.

The politics of the global warming debate have managed to make an objective analysis of the Holocene almost impossible though. The only factor that matters to many people now is what has happened in the past 50 years. It is absolutely impossible to make a reasonable climate assessment based on such a miniscule time scale. This is why I compare the Holocene to the Eemian.

Consider the warming that was needed to melt the Laurentide Ice Sheet that covered North America. This was an ice sheet that was comparable to Antarctica is size, but not in age. It contained enough water to change the global sea level by ~50m. The estimated volume was that it contained 30-35 million km^3 of ice. That number is meaningless in size, because it is just so large. Regular scuba divers can reach a depth of 30m which is still less than the amount of water that was stored in the Laurentide Ice Sheet. All of that melted in the early Holocene. Today, the climate is not warm enough to cause the Laurentide Ice Sheet to melt. It is that simple.

There is no better proof of long-term cooling in the NH than the fact that new glaciers are forming. The past 4,000 years has seen a resurgence of glaciers. They retreat at times and advance at others, but most of them didn't even exist 4,000 years ago and more often than they have been growing over that period of time. The story is clearest in the NH, but everywhere the story is the same. As the 65N insolation has dropped, so has the Earth's temperature.

The fact is that all the Northern Hemisphere ice cores show a steady drop in temperature. On Baffin Island north of Canada tells a similar story to Greenland. *Illustration 72* shows what the last 8,000 years look like there. I have added insolation to the chart to show that it has been dropping from its interglacial peak.

Greenland is more heavily impacted by changes in ocean currents than Baffin Island would be. That leaves Baffin Island more heavily dependent on insolation to control it's temperature. The ice core temperature responds much like expected with the temperature lagging the drop in insolation that is typical. Since this ice core ends in 1992 (8 YPB) it even shows fairly recent data. The recent peak would have been the 1917 point. Even the modern warming does not appear abnormal when compared to the longer term behavior. Again, the most recent 1,000 years are the **COLDEST** of the last 10,000 years.

I can show series after series that show cooling for the past 4,000 years in almost every part of the Earth. There are a few exceptions that show no cooling, but none of them are in the Northern Hemisphere.

Northern Canada:

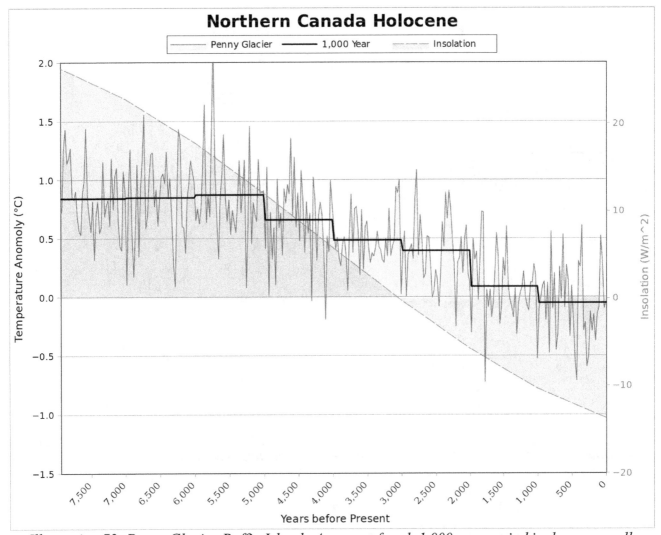

Illustration 72: Penny Glacier-Baffin Island. Average of each 1,000 year period is shown as well. Insolation has been dropping the past 10,000 years.

Baffin Island is north of Hudson Bay and it that contains the few remnants of the Laurentide Ice Sheet that did survive the Holocene. The trend there is the same as Greenland. The last 1,000 years are the coldest. The problem is finding ice cores that go back longer because most glaciers have only formed in the past couple thousand years because it was too warm before that.

It isn't just the NH though. Even Antarctica shows cooling over the past few thousand years. One very interesting ice core is the Taylor Ice Dome. Not only does it show significant cooling over the past several thousand years, but it also shows cyclical variation in temperature.

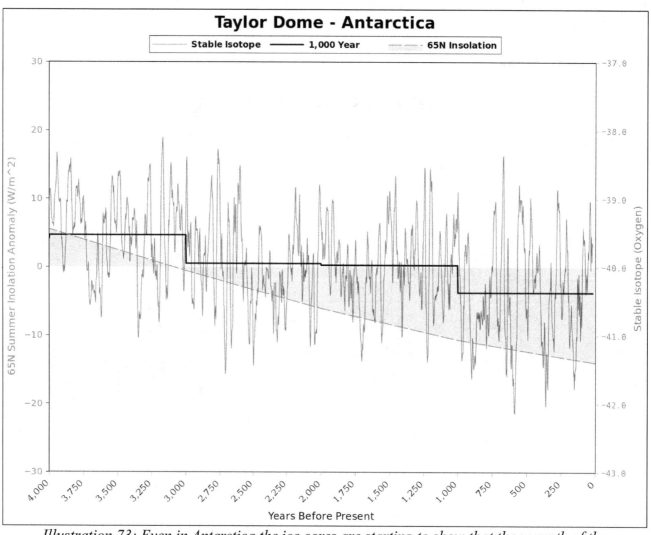

Illustration 73: Even in Antarctica the ice cores are starting to show that the warmth of the Holocene is fading.

Over the past 1,000 years the Taylor Dome ice core shows some significant variation in the stable isotope as the insolation has been dropping. The stable isotope for the Taylor dome shows that the Earth is much closer to a glacial period than it is to an interglacial. As with almost all other indicators, the long-term trend is cooling.

I would like to show also that the NH responds first in global cooling. I will do this by comparing the NGRIP (Northern Greenland) ice core to the EPICA (Antarctica) ice core from the Eemian. As the 65N insolation drops the NGRIP data shows a more rapid onset of cooling and the magnitude is also greater.

126

Eemian Temperature Anomaly

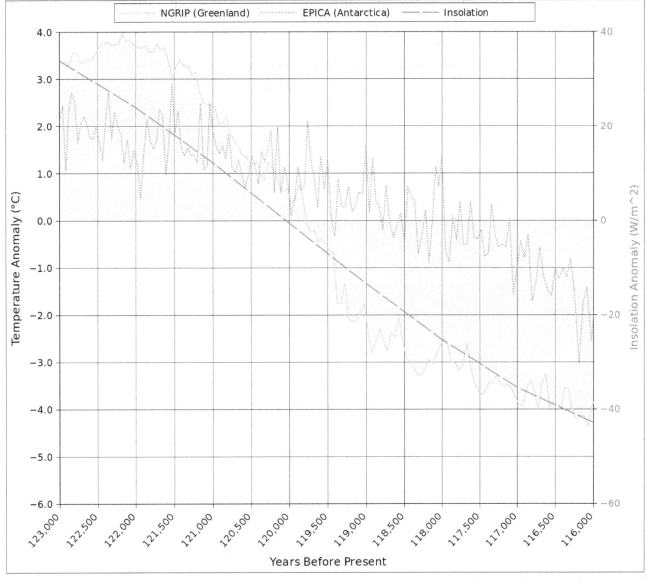

Illustration 74: NGRIP and EPICA ice cores from the Eemian. Insolation drops affect first the NH and then the SH. The rate of temperature change in the NH is greater as well than the SH.

This difference is important because in trying to predict the future, it is the NH where the change will be initially felt. So the NH will cool prior to the effects being measurable elsewhere. That also makes sense since it is the change in the NH climate that leads to the glacial effects.

Now the NGRIP[87] ice core is unique in that it is one of the only ice cores from the NH that spans back into the Eemian. Other ones experienced too much melting or movement in the past to provide useful data from the Eemian. The NGRIP ice core goes back clearly to 123,000 years ago. This makes it possible to compare the cooling of the Eemian to the current behavior of the Holocene.

So using a 7,000 year period from the Eemian and the past 7,000 years from the Holocene it is possible to compare interglacial to interglacial to see if the behavior is comparable from climate year to climate year. Even though the insolation drop is not as extreme during the Holocene as it was during the Eemian, the results speak for themselves.

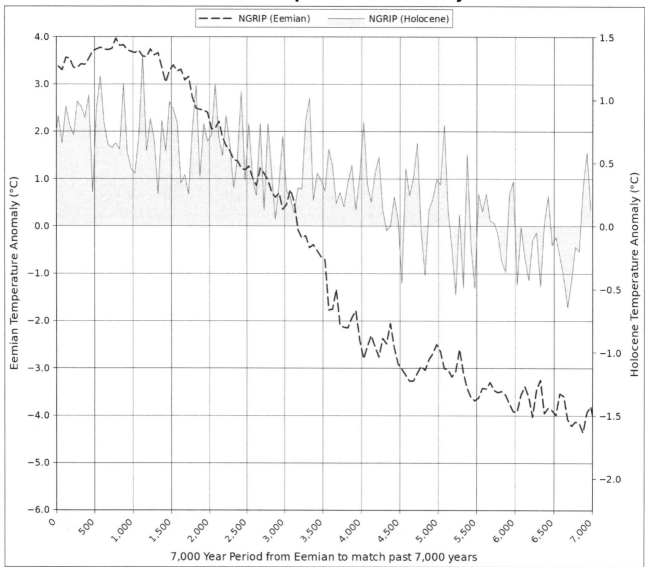

Illustration 75: The Eemian cooling took place more quickly, but the Holocene cooling is clear over the past 7,000 years in Greenland.

As with all the NH ice cores, the NGRIP shows that the last 1,000 years is the coldest of the past 10,000 years and it shows slow but steady cooling for the past 8,000 years. That rate of change is small, but the long-term rate of cooling always is a slow and gradual drop in temperature. Since the rate of insolation change in the late Holocene has also been more gradual than the Eemian, it is no surprise to see the Earth responding in the way that it is today.

The question you should be asking yourself now is: What is more significant, a short-term variation in the average temperature or the long-term trend that has been progressing for the past 3,000 years?

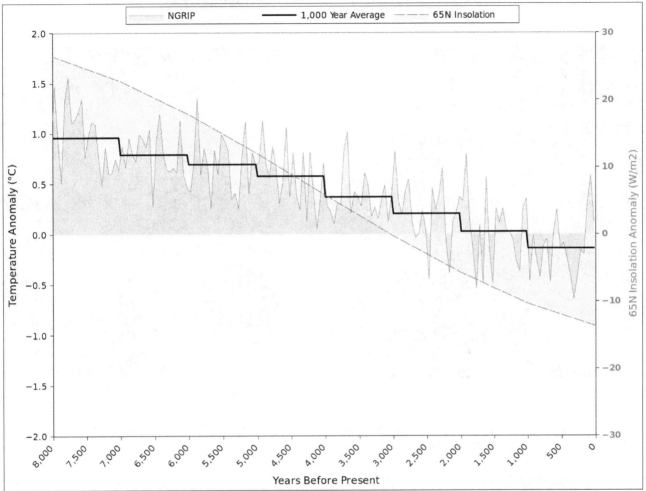

Illustration 76: NGRIP Ice core shows an 8,000 year cooling trend in Greenland.

Lost Cities of the Holocene

Anytime an underwater city is found the story of Atlantis pops up. Atlantis makes a nice story, but in this case reality is far more interesting. The following is a map where cities have been found at various depths under the oceans. The ones that date back more than 8,000 year are likely to be ones that were put underwater by the rising sea level of the early Holocene.

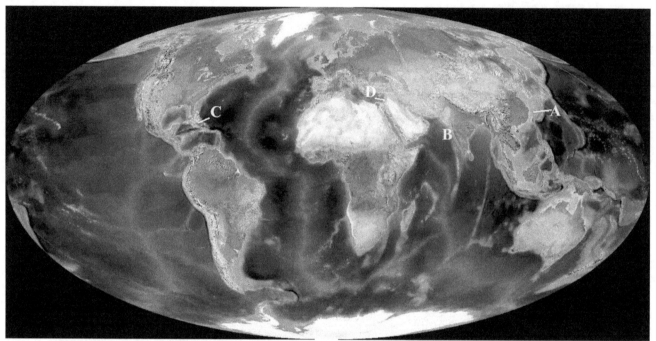

Illustration 77: Modern Map of the Earth by Dr. Ron Blakey, locations were added later.

A- **Yonaguni-Jima**[88]: Thought to be about 8,000 years old it has large complex that was found by a sport diver in 1995. There is significant debate as to the origin.

B- **Gulf of Cambay**[89]: In a pollution survey in 2001 a large city was found 36m (120ft) underwater. It is thought to be ~9,500 years old.

C- **Cuba**: The underwater city off of Cuba is very deep. There is still little known about this city, but the size and location make it possible that the Holocene contributed to its demise.

D- **Mediterranean**. Many cities are found off of Greece, Egypt and Israel. Most of the cities found so far are only a few thousand years old. The early Holocene had little to do with the sea levels then, but it is an indication that sea levels are always changing.

It is very probable that as people start looking for lost cities, more will be found in the future. It is likely that the real climate change that took place 15,000 years ago covered many cities when the oceans rose more than 100m.

Chapter 9   The Very Pesky and Inconvenient Warming and Cooling Periods

This is a chapter that should not exist. The fact that there is a need for this chapter is evidence that politics has taken over the global warming debate. Throughout the Earth's entire past climate, there has been constant temperature change in both the short-term and the long-term. Arguing that there has not been constant variation is purely a political game. Yet somehow, in the modern global warming debate, this has become one of the ugliest parts.

The entire argument about the warming and cooling periods is all about the short-term variation in the Earth's temperature over periods of 100-300 years. The argument is that there isn't normally significant warming or cooling in time frames of that scale. Okay, actually only a very select few people argue that the temperature doesn't fluctuate like that, but they are doggedly determined to prove that the Earth is naturally stable, unless the level of CO_2 is changing.

Much of the problem centers on the use of tree-ring proxies. Some argue that they provide a more accurate representation of climate and others disagree. Since these proxies are only useful for ~1,000 years (and limited in area over that short time period), I simply avoid using them. The issue is that some people have used tree-ring proxies to show that there is very little climate variability over the past 1,000 years. It is true that certain tree-ring proxies do not show significant temperature variation for much of the past 1,000 years. For some warmists that is enough to settle the science. What it shows to me is that tree-ring proxies are not very useful in detecting temperature changes.

I was not really aware of the how much fighting there was about the periodic warming periods until I was deep into my research. I had been aware that there was some disagreement, but the scope of the conflict was very surprising to me. Just mentioning the Medieval Warming Period (MWP) is enough to start a fight in some groups. In my view it should all be about the data as science is supposed to be about the facts. It was when I hit this topic that it struck me as to how much science has been abandoned in favor of politics.

On one side the warmists claim that the MWP did not really exist based on the tree-ring proxies. It was simply a localized warming event in the northern Atlantic region. They determinedly deny that such a period of local warming matters in the scale of things. There is even effort to reduce the size of the warming in places where the evidence is overwhelming that the warming was very significant.

Even the movie *An Inconvenient Truth* has a part where Al Gore made fun of skeptics for calling a little blip on his (claimed to be[90]) ice core chart the Medieval Warming Period. The thing he didn't tell you was that his chart wasn't an ice core chart at all. It was actually a reconstruction based on a tree-ring reconstruction. That all the ice cores disagree with the tree-ring reconstructions is soundly ignored.

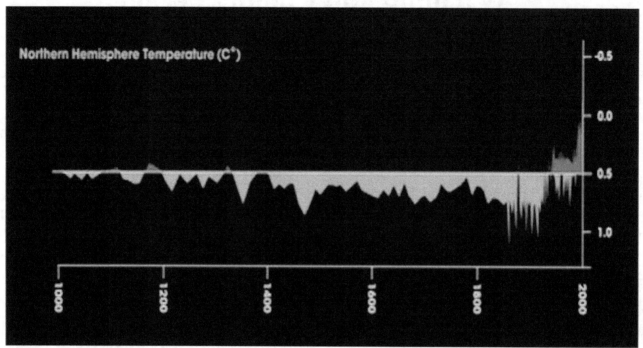

Illustration 78: This chart was used in "An Inconvenient Truth" to minimize the Medieval Warm Period. A careful reader will notice that in the modern time there is simultaneously warm and cold periods. This is because there are two sets of data overlaying. Neither one is from an ice core.

On the other side, the skeptics (self-proclaimed and otherwise) look at the tremendous amount of data that show the Roman and Medieval warming periods did exist in large portions of the world and are baffled by the statements that say otherwise. I can safely say that I was just as confused by those claims.

No one should be surprised by the fact that the Earth's temperature is always changing. The more one looks at long-term climate behavior, the clearer it becomes that stability is an illusion. Variation in the Earth's temperature has always existed and it always will. The problem for the warmists and the warming and cooling periods is one of perception. It weakens the argument that the current warming is caused by CO_2 emissions if the Earth was warmer 1,000 years ago. So minimizing previous warming cycles is a priority because that helps highlight the significance of the current warming period.

It is important to understand the background of this subject because any claims made about the two warming periods will be attacked regardless of their accuracy. There is no better example of how science has gone bad than the study of the previous warming and cooling periods. Facts have been replaced by opinion in many ways.

Since I don't view the previous warming periods as all that critical to what is going on in the climate, I will only touch upon them a little. Primarily I bring this up so you are aware of both sides of the conflict. In the end it doesn't matter because short-term trends have no predictive value for the Earth's climate.

Instead of arguing about how warm Greenland was 1,000 years ago compared to today, I will focus on comparing how much warming the Earth has experienced in the past 160 years compared to the amount of natural variation there has been in the ice cores over the past 2,000 years. I consider it normal for the Earth's temperature to swing up and down. To prove this I present 5 different ice cores: 2 from Antarctica and 3 from Greenland.

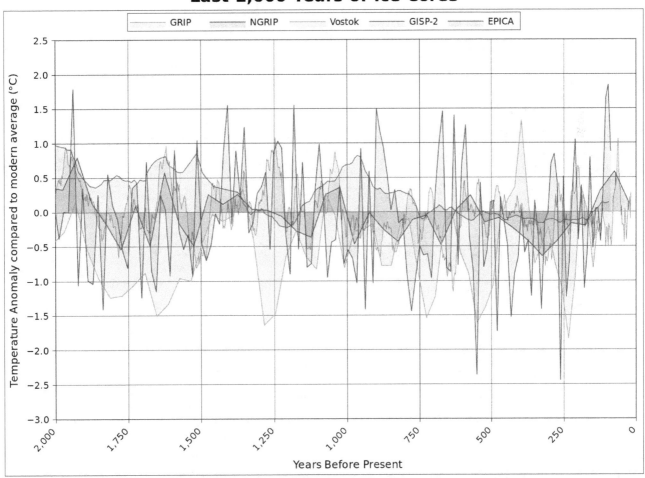

Illustration 79: (Antarctica) Vostok and EPICA, (Greenland) GRIP, NGRIP, GISP-2. All of them show significant natural variation in temperature. The purpose of this graph is to show the scale of natural variation.

All of the ice cores show positive temperatures ~1,900 YBP. Most then show a cooling trend for a few hundred years at which point all of them are warm again 1,300-1,400 YBP. There were periods where they were all cool again between 750 and 500 years ago. In the past 200 years they have generally been warm again. The range of temperatures for these ice cores is +/- 1.5 °C. This matches pretty much the type of behavior that is observed in most temperature proxies. The Earth's temperature has not been constant and the ice cores support that.

Now I can show how much the Earth had warmed up in the modern measurements and compare this to the ice core records. The results speak for themselves.

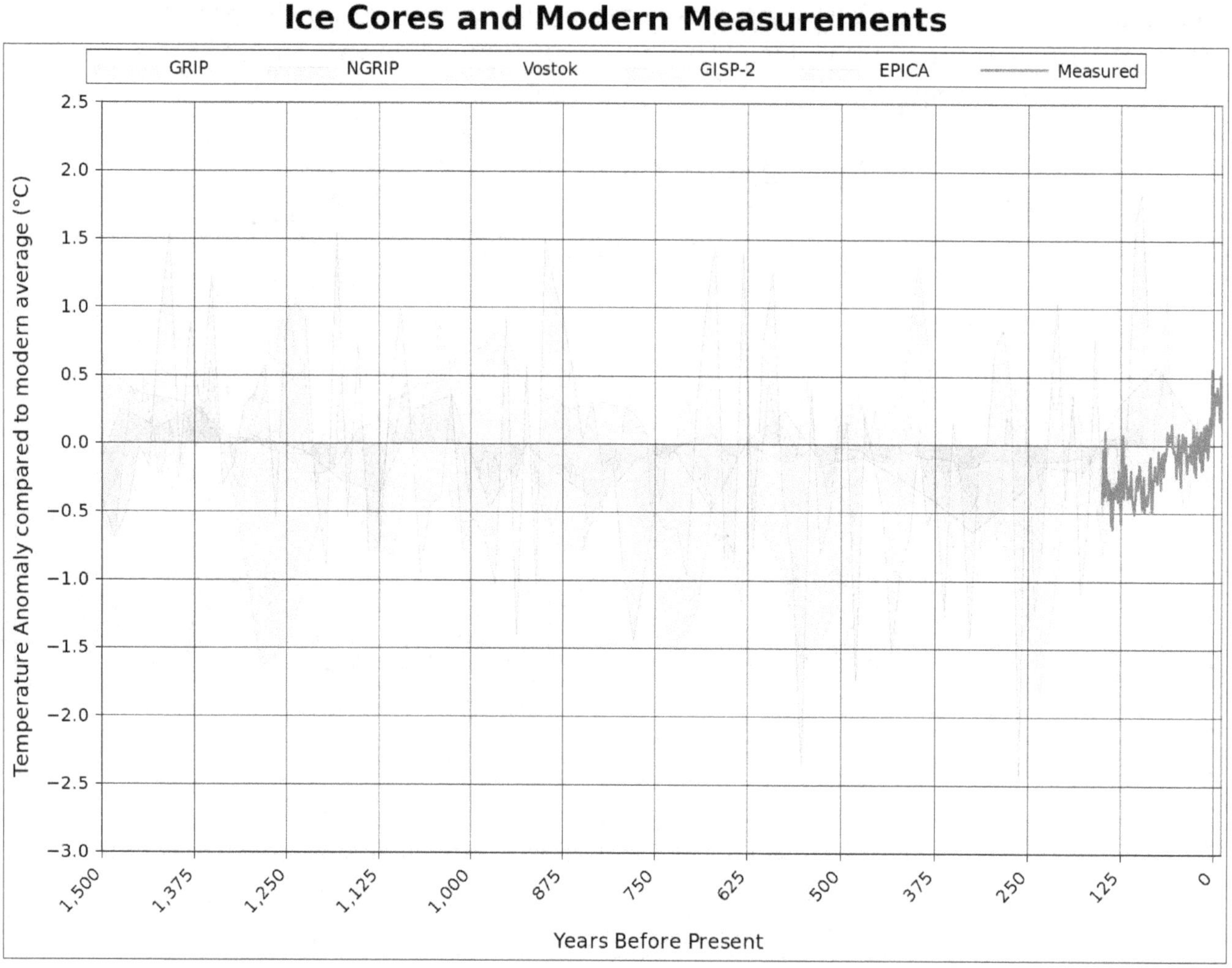

Illustration 80: Same chart as before, but with the modern instrument records overlaying the historic ice cores.

This is not an apples to apples comparison by any means, but it does give an indication of the type of scales that are being compared. It is also true that the Polar Regions are more impacted than the tropical regions, but once again, that has always been true. The tropics stay tropical, even during glacial periods, so the Polar Regions are the only valid indicator of "global" change.

I could pull hundreds of charts and write an entire book about the MWP, but there is no reason for such silliness. It is enough to know that the Polar Regions were much warmer 8,000 years ago and have been cooling for thousands of years. That is the perspective that I have about the Earth's climate. The Earth's temperature is right where I expect it to be for this part of the insolation cycle. The Earth's climate now is typical interglacial autumn. There are centuries that are warmer and centuries that are cooler. The whole idea that the Earth's temperature doesn't change is supremely erroneous.

Of course merely saying this isn't going to be enough. This is where the climate cycles show their usefulness. All readers will clearly realize that the only way to predict the future climate for December while in October is to go back to the previous year or the average of previous years. That is how the future temperature can be predicted. I would gladly lose the debate on the MWP for a warmist to concede that single point.

Once that point is conceded, then all that needs to be reviewed is the temperature behavior in the past climate cycles. If I define the interglacial autumn as beginning when the insolation anomaly goes negative, it is easy enough to use the EPICA ice core to show how the past interglacials behaved once the insolation anomaly went negative.

Interglacial Autumn

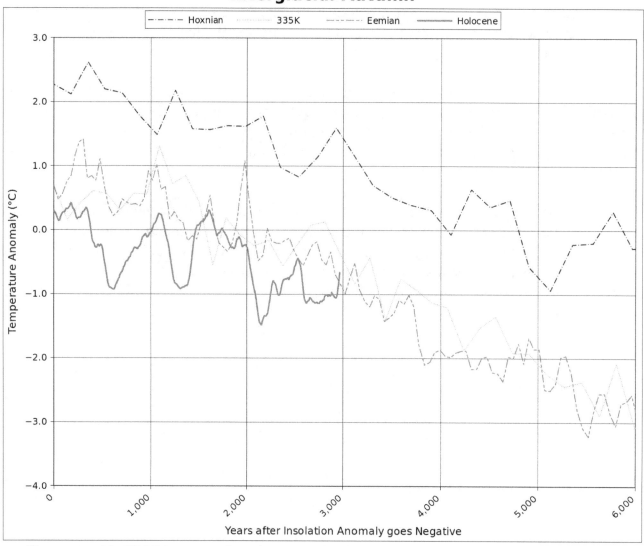

Illustration 81: EPICA ice core from four periods when the Earth was in an interglacial and the insolation anomaly went negative. In all cases the long-term trend in temperature was negative.

All of these are from the same ice core, but the Holocene did require smoothing. In every case the temperature of the Earth dropped once the insolation anomaly went negative. The Holocene is right on track with the other interglacials. The Hoxnian was an especially long and warm interglacial and it stands out as a very warm one in this chart as well.

This is typical interglacial fall weather. Some centuries are warmer than others, but the Earth is cooling down. Pick any peak and call it a warming period and pick any drop and call it a cooling period. That the last cooling period that took place was called the Little Ice Age in Europe should be sufficient warning that it was not the type of climate we want to live in.

Here are some of the climate highlights from the past couple of thousand years and how the climate impacted the people who lived at the time. None of the cold periods were good for mankind and all of the warm periods were. Pretending that the past did not happen will not help prepare for the future.

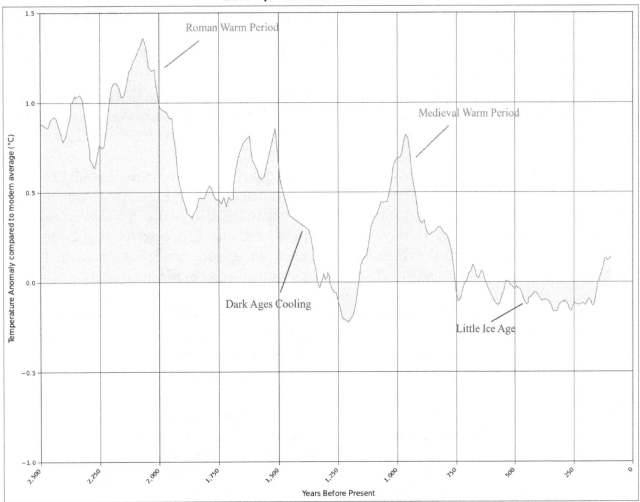

Illustration 82: This is a reconstruction from GISP-2 (Greenland). It does not show global temperatures, it shows the Arctic temperatures. Since that is the region that has shown the MOST variability over the past 2.6 million years, it is the region that matters most.

The Roman Warming Period (RWP) was a warming period that took place 2250-1500 YBP (250 BC to 500 AD). Documenting this from records at the time is difficult to say the least and many arguments erupt from references to wine made in England to the breadbasket of the Roman Empire being in North Africa. Scientifically precise information is simply not sufficient for most records to matter. So I will not get into that part of the argument.

One interesting fact is that the end of the RWP coincides with the onset of the Dark Ages in Europe. The climate did take a cold turn around ~500 AD. This period is associated with mass migrations and a return to a more rural lifestyle in Europe. Evidence of the colder climate exists in China where the capital was moved from Pingcheng to Luoyand in 494 AD[91] because even the summers were experiencing frost in the summer. That China and Europe both had migrations at the same time is an indication that a colder shift in the climate happened.

137

The cooling period lasted a couple of hundred years which happens to coincide with the main duration of the dark ages[92]. When that was ending 1,200 years ago happens to be when Europe started to warm up again into the main point of contention which is the Medieval Warming Period.

The MWP was a real event that lasted from about 800 - 1250 AD. There are certainly parts of the MWP that cannot be argued. The migration of the Vikings into Greenland and what is today Northern Canada[93] is strong evidence of warmer climate in those regions at that time. The regions simply cannot support farming today, but they did 1,000 years ago. In Greenland, the Vikings held on until ~600 years ago when the climate was simply too cold for their lifestyle to survive.

It is vehemently argued that this warming only took place in the Arctic regions and especially the Atlantic part of the Arctic. That is funny because that is the exact same region that is experiencing the majority of the warming today. Warmists call that region the canary in the coal mine for global warming. It is the Arctic that matters, but the description of the Viking colonization of Greenland and Newfoundland is exactly the description of an Arctic that used to be warmer. Then the whole region got colder over the course of a couple hundred years.

The Little Ice Age is that colder period that started as the MWP ended. The temperature of the Arctic dropped enough that farming in Greenland disappeared and only the Inuit people there survived the period of colder climate. It was not cold there in comparison to a glacial period, but it was colder compared to the MWP or the current warming period which I like to call the AGWP.

All of these are all small and mostly insignificant changes in the Earth's climate. None of them even come close to the scale of the changes that take place over the climate cycles. It is only to the people living in the places that are directly impacted that really notice these small changes. The Vikings expanded around the Arctic when the area warmed up 1,100 years ago. When Greenland cooled off 700 years ago, they died out.

There is perhaps one other reason why these warming and cooling periods are so inconvenient to the warmists: there was no variation in the level of atmospheric CO_2 that can explain such short-term changes in the Earth's temperature. The Antarctic ice cores[94] show that the level of CO_2 was stable during these periods, because of that they cannot be easily explained by the theory of global warming.

1,750 Years of Antarctic CO2 and Temperature

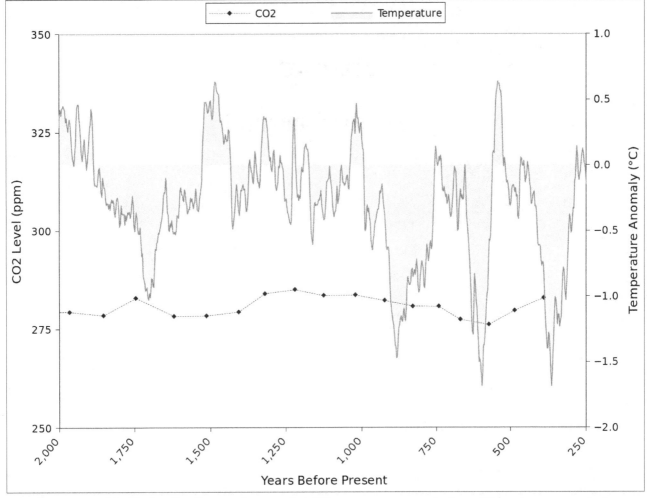

Illustration 83: Taylor Ice Dome shows +/- 2 °C temperature variation. There is no correlation between temperature and CO₂ level.

Even if those warming and cooling cycles only happened in the Arctic (unlikely if the above chart is correct), that is the region of the Earth that is currently driving the climate cycle. The only long-term trend that Greenland has experienced over the past 4,000 years is cooling. There have been several periods of a few hundred years that have been warmer than the trend. There have also been equally cold periods in that part of the world. That this pattern is continuing today should surprise no one.

Anyone who is fighting over the MWP is missing the big issue.

Summer is over

Fall is fading

Winter is coming......

Warning:

Scientific Content!!!

I try to keep the main chapters pretty level. If I need to rant I will do so in the Science Content. My daughter Sarah had to do a report on Greenland in 6ᵗʰ grade this year. One of the things she found was that 1,000 years ago the fishing in Greenland was different than it is today. The reason given in her history book was that the climate was warmer then. That led to a discussion about the whole warming thing.

When she asked her teacher at school why people were worried today when it was warmer then, the teacher didn't answer. I won't speculate as to why the teacher didn't answer because there are many possible reasons for a teacher to not answer. Even the most adamant of the warmists cannot deny that Greenland was warmer 1,000 years ago. They sure say the rest of the world was colder though.

Finding evidence of the MWP is easy. It took me only a few minutes to find impacts to the Mayan and Chinese cultures. That seems pretty world-wide to me. I am truly baffled and disturbed by the MWP deniers (I am smiling at that) ignoring obvious facts. Science requires an open mind. When I started my research I did not have a strong opinion about global warming. I was inclined to agree that it was possible because I thought I understood the greenhouse effect. Now it is clear that there is no scientific basis for global warming. Yet the debate continues.

I will put up some of the temperature reconstructions for around the Earth on the MWP. It would be easy to put together books analyzing the MWP. Entire books could be written about the MWP on each continent. It would not change many minds though. So I will show what the warmists say it was like and I will put up some opposing views just so you can see how crazy the fight over the MWP has gotten.

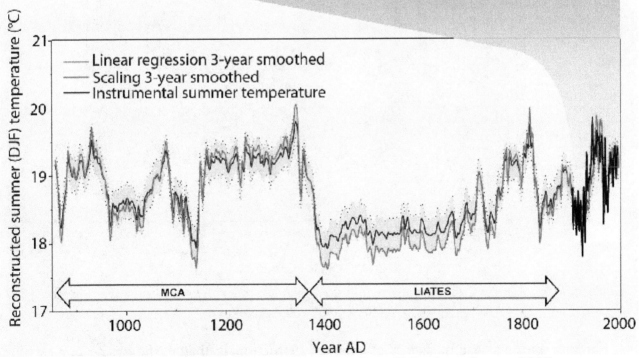

Illustration 84: Temperature Reconstruction from Chile. (von Gunten, 2009)

The von Guten[95] temperature reconstruction is fairly typical for South America. It shows that recent warming is happening, but it also shows that the region was also warmer in the past than it is today. This particular reconstruction is a summer temperature reconstruction. The limitation of such reconstructions is that they are limited to a particular region.

The 2005 reconstruction by Anders Moberg[96] is one that combined many sources for a reconstruction of the Northern Hemisphere. It also provides a value for each year which is helpful for analyzing the data. Since it combines different sources that used a variety of methods it is one that I prefer to use for the past 2,000 years.

2,000 Years of Temperature Change

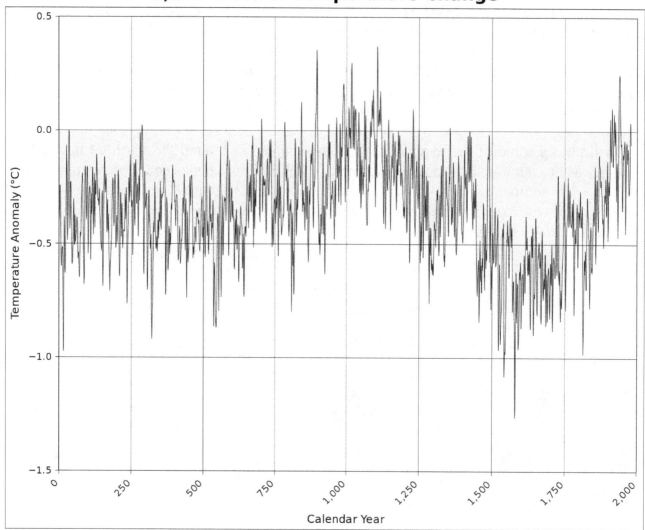

Illustration 85: Moberg 2005 Reconstruction (that includes tree rings) shows warmer climate 1,000 years ago and cooler climate between.

Perhaps the most useful aspect of this reconstruction is that it shows the coldest period happened ~400 years ago. From that coldest period of time, the NH started warming up. Since

142

there was not change in the atmosphere to cause the warming (or the cooling that preceded), there must have been another factor that caused both changes in the Earth's climate.

I usually use the 20 year average version as the data has less noise. Even using a moving average there is still significant variation that must be natural in origin as there is no reason to believe that mankind could have caused that variation that was taking place.

2,000 Years of Temperature Change

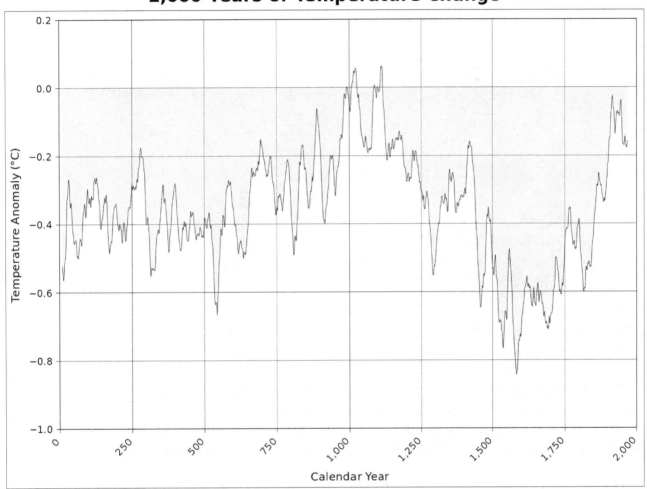

Illustration 86: 20 year average of the same Moberg chart.

The endless debate rages about how high the peak was 1,000 years ago compared to today. My only input is that it hardly matters because what this really shows is that the Earth's temperature is always in flux. It is not ever stable; it is always either warming or cooling.

Now compare the amount of year to year, decade to decade and century to century variation in the reconstructions that are most often used by warmists. They show very little variation at any time until the last 150 years. It is only then that the Earth seems to change much. The ones that don't show a MWP often look something like this:

Reconstructions of northern hemisphere temperature vary but all suggest it is warmer now than at any time in the past 1000 years

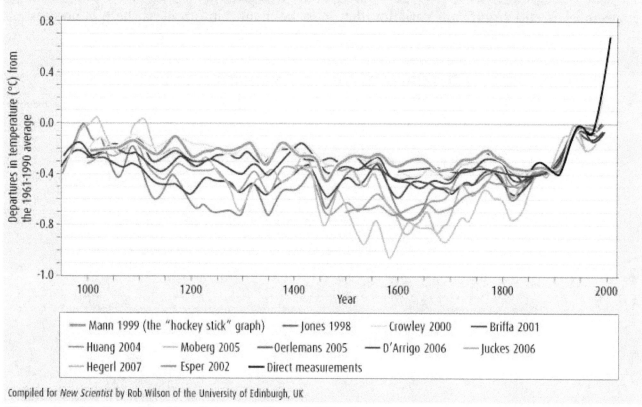

Compiled for *New Scientist* by Rob Wilson of the University of Edinburgh, UK

Illustration 87: From The New Scientist. Shows MWP almost as warm, but not quite as warm as today.
97

The same Moberg chart I used is in the mix here, if you look carefully. You can see that even more filtering is done to present such a smooth version of the Moberg chart in this graph. The 20 year average I used above still shows much more variation. This goes to show that with enough filtering and averaging, all usefulness can be removed from any source of data.

The "opposite" of Moberg is the Mann chart. That one also happens to be the basis for the chart used in the Al Gore movie. It shows the least variation over the past 1,000 years. The entire 1,000 year period fits neatly into a range of ~0.4 °C. Compare that to the Moberg range that is more than double at 0.9 °C. The 20-yr average has a range of 0.9 °C and the raw data has a range of 1.6 °C. If a person reviews enough data there is clear similarity in results by different scientists. The range of a reconstruction is just one way in which similarities show up.

In the end the only fact I need to know about the MWP is that the Earth was warmer than it was for most of the period that followed it. The reconstructions that show the Earth being colder support the simple fact I have been trying to make throughout the book. The last millennium has been the coldest one of the past ten. In almost every location around the world, each of the past four millenniums has been sequentially colder than the one before.

Arguing about the Medieval Warming Period in the face of that is pointless. There is enough natural variation to explain the current warming and later on I will even explain why there is so much natural swing in the Earth's temperature. The MWP might have been cooler in Tanzania than it is today, but that isn't relevant to what is going on.

The Earth has been cooling for thousands of years. I will not ignore that fact and neither should anybody else. Glaciers are now forming the NH halfway to the equator. Only 1,000 years ago some of those glaciers did not even exist. To me that is a more important climate indicator than the width of a few hundred tree-rings from around the world. Hopefully this chapter has provided a clear indicator of my views on the MWP.

Chapter 10 The Modern Warming Period (Aka AGW Period)

The summer of 1816 is known as "The Year Without Summer.[98]" The Little Ice Age still held sway over the Northern Hemisphere although the Earth had shown signs of warming over the past few decades[99]. That ended in 1815 when Mount Tambora[100] in Indonesia erupted which caused an almost immediate cooling as a result of volcanic ash and SO_2 that were emitted into the Earth's stratosphere. This blocked enough of the Sun's energy that significant cooling happened in the year that followed the eruption.

The United States did not suffer as much as other places because it was still not heavily populated. In New England many people abandoned their farms and went west in hopes of finding better farmland. Heavy frost in May that year killed many of the crops. New England experienced frost every month of the summer that year[101]. In early June there was a cold front that killed off the crops that frost had not killed and caused snow to fall in New York on June 6, 1816 [102]. Ice was found on lakes and rivers throughout the summer that year. If it was not for the crops in the southern parts of the United States the effects could have been much worse.

Europe does not have a warmer southern region and it also had a larger population. This made Europe more sensitive to the effect of cooling than the United States. Hundreds of thousands of people died across Europe from a combination of starvation and the violence of food riots. Heavy frosts in August devastated what few crops had survived that long. It was one of the worst food shortages in the modern western world[103].

The NH experienced a disaster that year. Food was scarce and even when it could be purchased the prices had skyrocketed. One year without a harvest resulted in disaster throughout the modern world. In those days, long range transportation systems for most foods were non-existent. Communities depended on what they could grow.

Perhaps the world today could cope more effectively with a comparable disaster, but there are far more people in the world today that do not grow any food at all. This also ignores the fact that much of the food that is grown today is converted into fuel for vehicles[104]. Comparisons aside, such an event today would cause food prices to escalate dramatically. Even if society today could survive a 50% crop failure for single year, the impact of such an event would be traumatic on society.

"Civilization is only two meals away from anarchy." *unknown, but possibly Larry Niven*

In the end there are only two possible futures for the Earth's climate. If CO_2 causes warming, then the Earth *might* possibly miss the next glacial period that has been developing for the past few thousand years. If CO_2 does not cause warming, then there is no question that the weakening Holocene Interglacial will end in the next few thousand years. The one future that does not exist is one where the Earth's climate will become stable and stop changing. The climate will change, it always has and it always will.

This is why I always refer to the theory of global warming as global warming. Climate change is the natural and constant state of the Earth. It should be very clear that at this point in the Earth's natural cycle, the Earth should be slowly cooling into a long glacial period. The next glacial period will likely last until ~105,000 years in the future. There will be a couple of periods in there where the Earth will warm up, much like what happened during the last glacial, but that is the natural direction of the Earth today.

The real question people should be asking is; can CO_2 alter that future? When Arrhenius first proposed the idea that CO_2 regulated the ice age cycle[105], he and others wanted to use CO_2 to warm the Earth up as they thought it would be a better world if it were warmer. In that regard they were correct. Glacials are the most dangerous parts of the climate cycle. Agriculture only came into existence after the Earth entered the Holocene. Glacial periods are not good for agriculture.

So far I have focused on explaining the cause of the natural cycles of the Earth. Geography and solar insolation are the driving forces in the climate changes that have taken place over the past 50-60 million years and there is no reason to believe that those two items did not determine the climate prior to that, nor is there a reason to believe that those two factors will not continue to determine the behavior of the Earth's climate.

The problem with those two driving forces is that they both act in the long-term. They change slowly and the effect that they have on climate can only be measured over thousands of years. In time periods of a few hundred years they do not change enough to have much of an influence on the climate. The change in 65N summer insolation in the past 200 years is negligible at -0.15%. A change that small does not cause significant changes in the climate much like the difference in insolation between July 10th and July 20th does not make much of a difference.

This means that other factors must play a role in causing the short-term changes in the Earth's temperature. These factors have little effect in the long-term, but they do cause some of the changes that take place on time scales that people can observe in their lifetime. I do not propose that they are solely responsible for causing the warming that has been observed over the past 160 years, but they have played a role in changing the Earth's temperature at different times in the past 160 years.

It is also important to distinguish here between temperature and temperature anomaly. As the book proceeds, the actual temperature instead of the temperature anomaly will play an increasingly important role. While many people tend to focus solely on the temperature anomaly, the actual temperature of the Earth is in many ways more important than the temperature anomaly.

HadCRUT3 Global Temperature

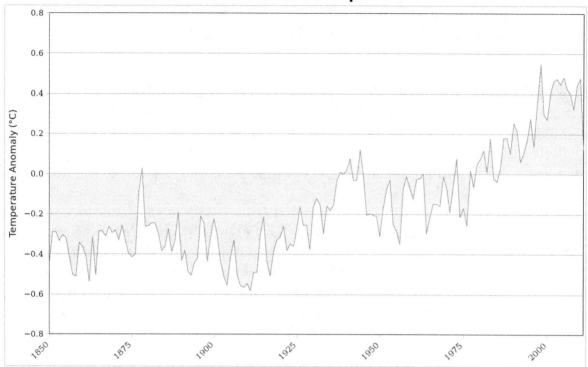

Illustration 88: The HadCRUT3 is the temperature anomaly for the land and oceans as measured at locations at the Earth's surface

This chart is very typical of the type used to 'prove' global warming. It shows warming between 1930-1945 and again 1980-2000. The end result is that the decade from 2000-2010 is 0.8 °C warmer than the decade from 1850-1860. That certainly gives the appearance of a warming trend. The facts are more complicated than that.

Now I am going to show the Earth's temperature and the temperature anomaly on the same chart for the period from 1979-2011. Instead of showing only the yearly data, I am going to show the temperature and the anomaly for each month.

The biggest difference is the scale for the temperature anomaly. I have made it match the temperature scale that shows how the Earth changes over the course of the year with each scale being +/- 2.5 °C. That is proper scale for showing the annual variation in the Earth's temperature. It also shows that the 'warming' is small in scale compared to the annual change in the Earth's temperature.

Monthly Global Temperature

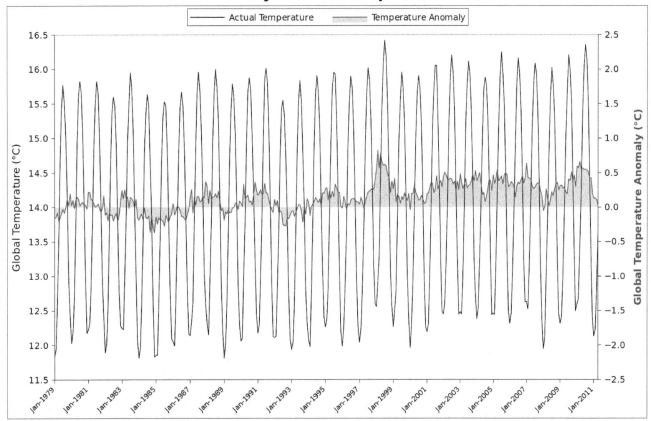

Illustration 89: (Black) Actual Global Temperature (Purple) Temperature Anomaly since 1979.

Even over the course of a few years there is a natural back and forth between warming and cooling of the temperature anomaly. That alternating warming and cooling is part of the natural variation that the Earth constantly goes through. It is that variation that causes each year to have slightly different weather. Only by understanding the causes of the natural variation is it possible to determine if something unnatural is going on.

The most common cause of the natural variation is the El Nino / La Nina phenomenon that occurs in the Pacific Ocean. The scientific name for this is the El Nino Southern Oscillation (ENSO). As the name implies it is an oscillation in the sea surface temperature of the equatorial Pacific Ocean. During an El Nino the water warms up and during the La Nina the water cools down.

The ENSO has an enormous impact on the temperature of the Earth's tropics. In effect, the ENSO causes 33% of the very short-term variation in the Earth's temperature. Most of the peaks and valleys in the temperature anomaly charts are caused by the ENSO. Note that the ENSO leads the temperature in the same manner that insolation leads the seasonal temperature changes. The typical delay for the atmosphere is ~5 months. When the ENSO starts happening, it is fairly easy to predict what the temperature will be doing 6-10 months later.

Tropics: Satellite Anomaly and ENSO

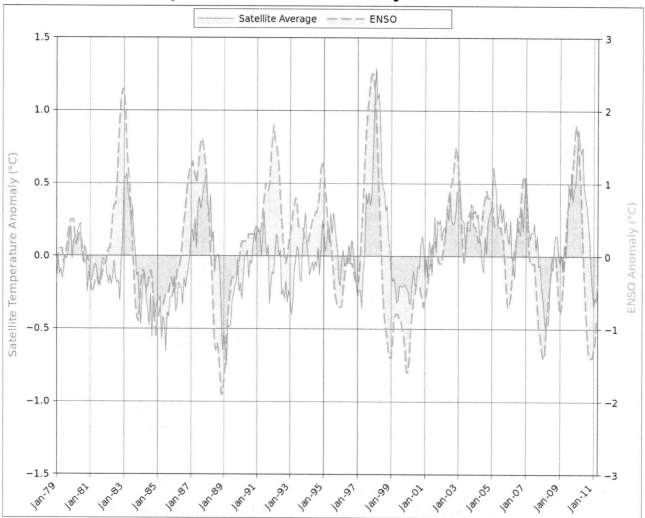

Illustration 90: Tropical temperature anomaly and the ENSO.

It is unknown what causes the ENSO, but it is known that the ENSO has been happening for a long time. A reconstruction by Dr. Cobb[106] showed that it has been happening steadily for the past 1,100 years with the most intense ENSO activity taking place in the mid-seventeenth century. This proves that the ENSO has not been caused by human activity. There is also no question that the ENSO does cause the Earth's temperature to warm and cool depending on which phase of the cycle it is in.

There is one period on the above chart where the temperature in the Tropics does not follow the ENSO. That period starts in June of 1991 and continues until the end of 1994. For that 2.5 year period the temperature of the Earth did not react in the normal manner to the ocean temperatures. The ENSO went very positive in that period of time and the temperature of the atmosphere dropped. That disruption in the normal behavior was caused by the massive eruption Mt. Pinatubo. Most volcanoes do not have any noticeable impact on the Earth's climate. Eruptions happen frequently; generally there are at least a few that erupt each year. It is rare

150

when an eruption has an effect on the Earth's temperature, but Mt. Pinatubo was one of the massive volcanoes that sent large quantities of ash and SO_2 into the Earth's stratosphere. When those materials get into the stratosphere in any quantity there is a measurable impact on the Earth's climate.

When Mt. Pinatubo erupted on June 15th, 1991 it put millions of tons of ash and SO_2 into the stratosphere. In the lower troposphere, it quickly gets washed out of the atmosphere by clouds and precipitation. In the stratosphere it takes years for the material to dissipate. It takes ~6 months for the ash and gas to migrate around the entire Earth. Once it does they reflect sunlight away from the Earth. This reduces the insolation that reaches the Earth's surface and causes the Earth to cool down. One year after the eruption the NH had cooled down by 1 °C.

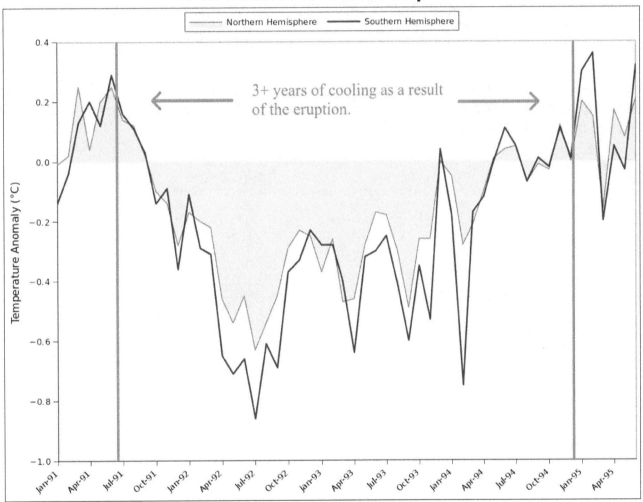

Illustration 91: UAH satellite anomaly for each hemisphere for the period around the eruption. Such an event had an obvious impact on the Earth's temperature for years after the eruption.

In the past two hundred years there have been at least 3 volcanic eruptions that were large enough to cause measurable cooling of the Earth. Mt. Pinatubo and Mt. Tambora have both been introduced. The other volcano to cause significant cooling was the eruption of Krakatoa in 1883.

This was in-between Tambora and Pinatubo in strength. It is often stated that it took five years for the weather to return to normal after that eruption.

These two natural sources of temperature variability are very short-term. Even the most powerful volcanic eruptions of the past 200 years do not have an impact on climate for more than 10 years and those always cause cooling. The ENSO can cause warming and cooling, but only for a year or two at the most. Both of these are evidence that there is always going to be variation in temperature, but they do not explain temperature changes that take place over decades or centuries.

There are two other natural ocean cycles that are currently known to cause some of the natural variation in the Earth's temperature. These are the Pacific Decadal Oscillation (PDO)[107] and the Atlantic Multidecadal Oscillation (AMO)[108]. Both of these are less well known than the ENSO and their effect is less dramatic. This is balanced by the fact that they tend to swing back and forth more slowly. The PDO tends to change on 10-20 year periods. The AMO takes at least 70 years to complete a single cycle.

The AMO is especially interesting in the global warming debate because it has a profound impact on the Arctic. Greenland and the Arctic sea ice are the two areas that seem to be heavily impacted by the different phases of the AMO.

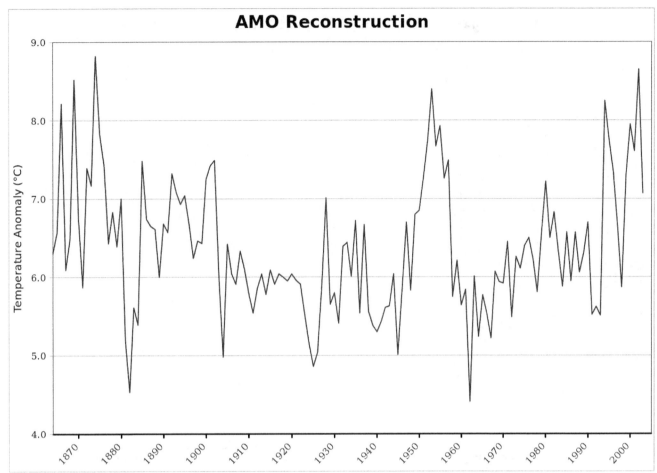

Illustration 92: 139 year reconstruction of the Atlantic Multidecadal Oscillation.

This supports that there has been a short-term warming trend since 1970, but it is certainly nothing that stands out as extraordinary. The same study[109] that did the above reconstruction included temperature reconstructions for the same region from the 10th and 13th centuries. The Milankovitch expectation for a region in the NH over a 1,000 year period would be cooling.

That is exactly what the AMO reconstruction shows. Each period of the AMO is cooler than the one before with the latest one being the coldest of all. Even the factors that cause short-term variability in the Earth's temperature show signs of long-term cooling. This is of course the expected behavior for this season of the climate cycle.

1,000 years of the Atlantic Temperatures

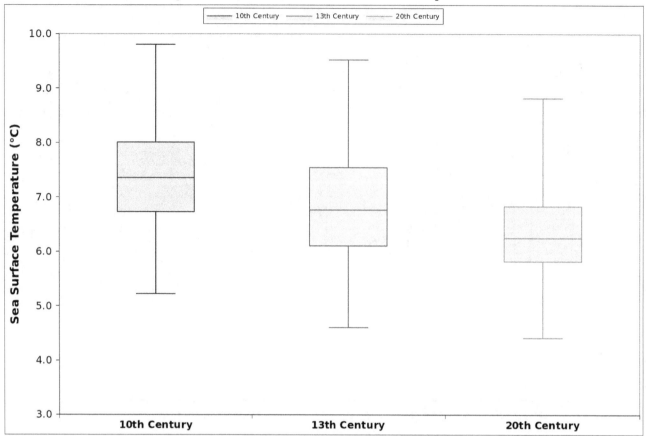

Illustration 93: The AMO over the past 1,000 years. The average, upper limit and lower limit are all trending down for the past 1,000 years.

The sunspot cycle is another potential source for short-term climate variability. This is one area where I do not find the evidence sufficient to reach a conclusion. Many other people consider that a likely source of the recent warming, but I am not convinced that this is so. I have written more about this on my website[110] if you are interested in why I have reached this conclusion. I write this only so people are aware that I am intentionally leaving it out because the data for this theory is not sufficiently convincing to me.

These sources of variability do explain some of the warming that has taken place in the past 40 years. I do not believe they explain all the warming that has taken place. There is one final source of variability that I believe is the key regulator of the Earth's climate. That final source would be sufficient to explain the recent warming and even more warming that might take place over the next 50 years, but not much beyond that. This final source will have a later chapter dedicated to it, so I will leave it until then.

These are reasons why the Earth's climate has a certain amount of variability. There are other proposed causes and some of them might have an impact, but these are the ones that I have been able to find that show a clear cause and effect. In each of the cases that I discussed the cycle or event is known to be natural.

154

Looking at the all-time high record temperatures is another interesting way to compare the present day (last 10 years) to the past 150 years. The United States has one of the largest and most extensive sets of temperature data. As a result it has accurate temperature records for many locations that date back more than 100 years in many places. The hottest recorded temperature for a particular state is a very practical way to see how hot places are getting when compared to the past.

The high temperature for a particular day of year is not the type of record that I am talking about. Records like that are broken almost daily[111] for high and low temperatures. What I am talking about the highest temperature ever recorded for a state. That the United States happens to have an extensive set of all-time high temperatures for each state makes it an easy way to compare the warmth of the past 10 years to the past. Of course this only applies to summer temperatures because by default the high temperatures will always take place during the summer[112]. Since warmists like to say that the Earth is warmer now than it has been in a 1,000 years, it would make sense that all-time records are being broken every year.

When I put this to the test, the results were very surprising. The last time a state set an all-time high record temperature was 1995. Connecticut managed to set an all-time high record of 106 °F (41.1 °C) back on July 15th of that year. In 1994 there was a hot spell that broke all-time highs over a 3-day period in Arizona, Nevada, New Mexico, Oklahoma and Texas in late June. Prior to that you have to go back to 1983 to find an all-time high temperature record in the United States.

What is really interesting is that 25 states, half of all states have all-time record high temperatures from 1930 – 1937. That seven year period holds half of the all-time record high temperatures for the United States. Even more curious is that only 15 states have all-time record highs that are more recent than 1937. It has been 16 years since any state has set an all-time record high temperature. The average all-time record high temperature in the United States is ~70 years old.

1936 was a very hot year for the United States. A total of 14 states have record highs from that year. It wasn't just one heat wave that set those records either. They are scattered from July 5th to August 10th, although July 10th does have 4 of the records for the states of Maryland, New Jersey, Pennsylvania and West Virginia.

The last time a state set an all-time record low temperature in the United States was February 11, 2011[113]. So it has been 16 years since a record high temperature was set and several months since a record low was set. Once again I am struck by the oddity that warming is what I should be concerned about.

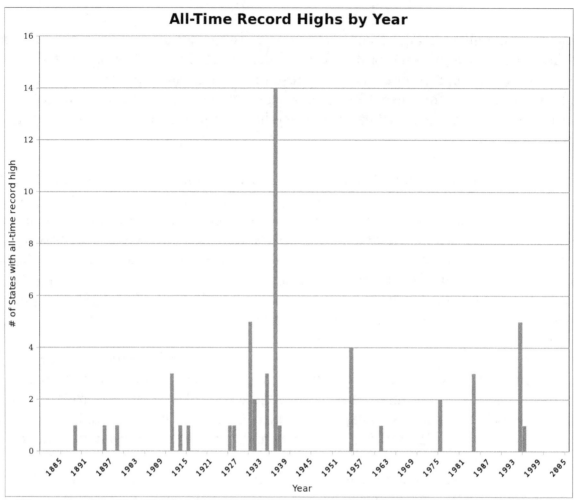

Illustration 94: The 1930's have 50% of the All-time High Records. High's only happen during the summer months. The big peak in 1994 was a very hot day in the southwest that caused records on a single day. In the 1930's the records were scattered across the entire summer. None of the all-time high temperatures have happened in the past decade.

According to the Milankovitch Theory of the climate cycle, the summer temperatures should be dropping as a result in the less direct insolation from the sun that is happening. If the all-time high record temperatures are an indication, then the peak summer temperatures have not been increasing over the past 70 years. That 68% of the states have record highs from the dates between July 11th (Colorado) of 1888 to July 5th of 1937 (Montana) is an indication that peak summer temperatures are not increasing.

It is cooler summers that cause glacials to develop. The average temperature doesn't cause anything to happen. It is all about when and where the Earth is warm and cool. As time goes forward the winters in the NH will continue to warm, but the summers will not. Until the glaciers grow large enough that they keep even the summers cool. Then the NH temperature will plunge until summer insolation rises once again. The trade-off is always happening because decreasing summer insolation is always balanced by increasing winter insolation.

156

Based on the all-time record high temperatures in the United States, it is clear that the summers in the past had higher peak temperatures than we are experiencing today. Some of those record temperatures may last a very long time if the Milankovitch Cycle progresses as it should.

Of course that is just the United States. What about the record high temperature for each continent? Certainly these should show global warming if it were happening? The average age of the all-time record high for each continent is currently 92 years old[2]. That is even older than the average all-time record high temperature in the United States. This is even if I use the 1960 record for Australia, the next oldest record in 1942. 5 of the 7 continents (and Oceania) have records that are ~100 years old[114]. It would appear that maximum temperatures have not increased anywhere on the planet Earth.

---

2   Excluding Antarctica as reliable temperature records only started in 1955.

The highest temperatures for each continent/region also show that the high temperatures are not increasing.

Locator #	Continent	Highest Temp. (deg F)	Place	Elevation (Feet)	Date
1	Africa	136	El Azizia, Libya	367	13 Sep 1922
2	North America	134	Death Valley, CA (Greenland Ranch)	-178	10 Jul 1913
3	Asia	129	Tirat Tsvi, Israel	-722	22 Jun 1942
4	Australia	128*	Cloncurry, Queensland	622	16 Jan 1889
5	Europe	122	Seville, Spain	26	4 Aug 1881
6	South America	120	Rivadavia, Argentina	676	11 Dec 1905
7	Oceania	108	Tuguegarao, Philippines	72	29 Apr 1912
8	Antarctica	59	Vanda Station, Scott Coast	49	5 Jan 1974

Highest Temperature Extremes

```
* Note: This temperature was measured using the techniques available at the time of
recording, which are different to the standard techniques currently used in Australia. The
most likely Australian high-temperature record using standard equipment is an observation
of 50.7°C (123°F) recorded at Oodnadatta in 1960.
```

The claim is that the Earth is getting warmer, but most of the all-time record high temperatures are already older than the average human lifetime. Part of the issue is that average temperatures are just limited in the information that they provide. The average temperature of the past 10 years might be warmer than the 1930's, but the summers have not been as brutally hot in the past 10 years as they were in the 1930's.

Another way to look at this is of course, the climate cycle. In the United States, 49 of the 50 states all-time record high temperatures were set between June 27th and August 21st and most of them are in July (Hawaii was in April, but the behavior there is different). Setting an all-time record high temperature after September has arrived is almost impossible simply because there isn't enough energy from the Sun. The Earth is now farther into the climate cycle than it was 100 years ago. While the difference in insolation over 100 years is not large, setting all-time record temperatures takes a lot of energy. The older the record temperatures get (especially in the NH), the less likely it is that they will ever be broken, at least in this climate cycle.

The other factor is that while the 65N summer insolation is dropping, the winter insolation is increasing. The correct long-term trend for the Earth today based on the Milankovitch Cycle is decreasing summer temperatures and increasing winter temperatures. This can be compared to the rates of warming and the result is that the winter rate of warming in the NH is double that of the summer warming. When the observations fit the theory, that is usually evidence that the theory is correct. The only theory that fully explains the climate cycle is the Milankovitch Cycle.

All indications are that the small variability of the **average** temperature of the past 150 years is no more significant than an above average week in October. The temperature is not summer-time hot, but it is warm for October. That is where the Earth is now. It is experiencing an Indian Summer[115] in the climate cycle. A weekly warming trend in October is not proof that winter will not arrive. An Indian Summer is also a warming trend against the insolation trend, but never has one been used to claim that winter will not arrive. A person claiming that an Indian Summer meant winter would not arrive would be promptly ignored.

Indian Summer

In North America an Indian Summer is when the weather turns warm in the autumn, usually after a frost or other cold spell. Usually the temperature will go up at least over 20 °C (70 °F) for a few days or sometimes as long as a week.

In Western Europe the same thing is called a St. Martin's Summer. Eastern Europe will call it a Women's summer, Gypsy Summer or several other things depending on the country.

Regardless of what it is called, the phenomenon is common enough that people understand what it means. It doesn't happen every year and sometimes it is very short lived. They show up in records for a very long time throughout Europe so there is no reason to believe that they have not always happened.

Such warming events happen in the climate cycle as well. The long-term trend in the NH for the past 4,000 years has been cooling, but there have been 5 periods of significant warming in that time. The current one fits right in with the last 4.

If the Earth is warmer now than any time in the past 1,000 years, then why have none of the all-time record high temperatures been set for any inhabited continent in more than 50 years? Why are the record high temperatures almost 100 years old with no end in sight?

Today (May 28th, 2011) I read that Aspen[116], Colorado is going to open the ski lifts because almost two months after closing, there are still 70 inches (1.8m) of snow at the base. That is not how the climate of Aspen has behaved in the past. There is more snow later in the year than usual. There is certainly an aspect of weather variability in that, but that is exactly how glaciers form. The snow fails to melt during the summer and the next year there is more snow as a result. That first layer that fails to melt is the most important part in the formation of a glacier.

For the past 4,000 years there has been a gradual cooling of the Northern Hemisphere. Glaciers have started to form again as the summers have received less energy from the sun each summer. Most of the glaciers in the NH are less than 4,000 years old. There have been many warm periods in the past 4,000 years, but increasingly over the past 1,000 years there have been longer and colder periods, especially in the NH.

I know I have said this before, but comparing the interglacial summers and falls is the only way to be able to accurately predict what is going to happen next. Just like comparing the fall temperature behavior year after year is the only way to understand how fall behaves. All I have done is look back far enough so the past autumns can be compared to each other. Much like the beginning of the annual autumn is defined by the day the Earth gets even energy from the sun; the interglacial autumn appears to be defined by the suns insolation dropping below the average.

All of this is what the Milankovitch Theory predicts. It will be 25,000 years before the NH once again receives a positive insolation anomaly. It will be nothing like the insolation peak that caused the Holocene; instead it will be a much weaker version of the flat insolation that caused the extreme climate instability 30,000 years ago. The last thing that I am worried about is an Earth that is warmer than it is today. What I worry about is one that gets colder than it is today.

It will be ~105,000 years in the future before the same conditions exist that led to the Holocene Interglacial. Until then it will be far more like the conditions that existed between the Eemian and Holocene Interglacials. The oceans will be lower, ice will cover much of the Northern Hemisphere and life on Earth will be much more difficult.

There are only two possible climate futures. The natural one is where a full glacial develops over the next couple thousand years and civilization as we know it today will no longer be possible. The alternative is that CO_2 emissions have altered the climate enough to prevent the natural future from happening. The Earth will not stay the same. That future does not exist.

Warning:

Scientific Content!!!

Indian Summer

Here is the daily temperature for Manhattan, KS from the fall of 2010 to spring of 2011. It shows a 3 day Indian Summer in early November. How useful are those 3 days in predicting the future behavior of the temperature?

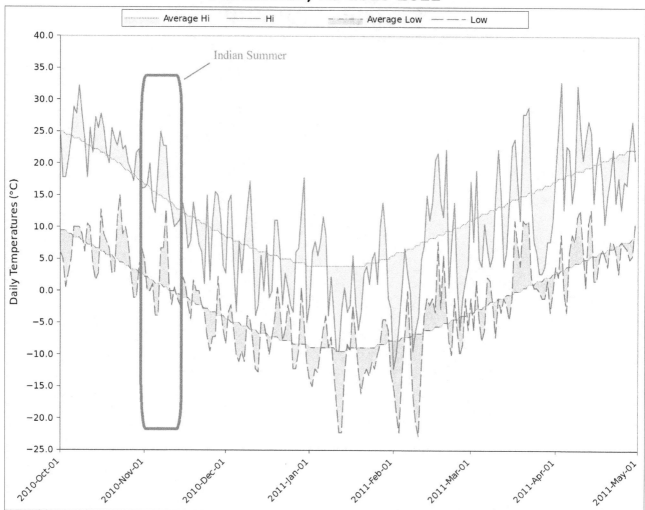

Illustration 95: The circled section in November is an Indian Summer. The temperature prior to that had been freezing during the night. Then the temperature increased to day time highs well over 20 °C and lows ~10 °C. That warming up against the natural trend meant little a month later when high temperatures were below 0 °C.

Why do Indian Summers happen? There is no great mystery there, weather happens. A warm front will move in from somewhere and that warm air behind it will have above average temperatures for those few days. In Kansas, the odds are just as good that any given Fall day will experience below average temperatures. It is truly a case of variability.

When an Indian Summer happens, no claims are made that winter will not arrive; no cause and

162

effect analysis is needed. People understand that weather is unpredictable. The absolute worst reason to argue that global warming is real is because the Earth has warmed up. Such an approach ignores natural variability and leads to an attempt to find a cause. Warmists then seize upon CO_2 trends to 'prove' that emissions have caused the warming because nothing else could have caused the warming. They completely ignore what the one chart above shows. Only 29 out of the 212 days have a temperature anomaly of +/- 1 °C. Only 14% of the days can be described as average.

Anomaly and Temperature

In climate science terms the above chart would be boiled down to a -0.93 °C temperature anomaly. What information does that provide? A temperature anomaly by itself has very little meaning. Here is the information that I prefer to look at.

Average High: 12.4 °C   High Anomaly: -1.4 °C

Average Low: -1.5 °C   Low Anomaly: -0.5 °C

This provides much more detail about what is going on. If such information was readily available for the different hemispheres and seasons it would be much easier to see what the Earth's climate is really doing.

The difference between the high and low temperatures is the diurnal temperature range. Places that have large differences will generally have larger temperature anomalies over time, although there are factors (large bodies of water and humidity) that can alter that behavior. The same general observation that the more temperatures change, the greater the anomaly is.

One interesting example is the Earth's two hemispheres. This is also where actual temperature starts to become important. Some basic temperature information for the two hemispheres is:

Northern Hemisphere: 14.5 +/- 7.1 °C. Anomaly range: 1.6 °C

Southern Hemisphere: 13.2 +/- 3.3 °C. Anomaly range: 1.3 °C

The NH as a whole has more than double the temperature change between the winter and summer seasons. It should be no surprise at all that it sees larger temperature anomalies as well. It also has a higher standard deviation in the temperature anomaly.

To put this into chart form, I am going to show the hemispheric temperatures and anomalies, but using the exact same temperature scale for both. This shows just how small the anomaly is when compared to how much each hemisphere changes over the course of a year. I did this by applying the UAH anomaly to the NOAA global temperature data[117]. The UAH was picked only because it provides the most comprehensive spatial coverage.

Northern Hemisphere Behavior

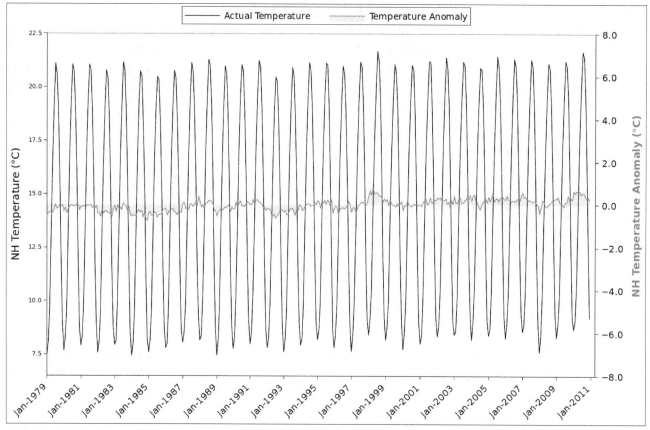

Illustration 96: The NH gets a +/- 8 °C range. The temperature anomaly for that range is miniscule, but it is the NH anomaly that drives the global temperature anomaly in most years.

Southern Hemisphere Behavior

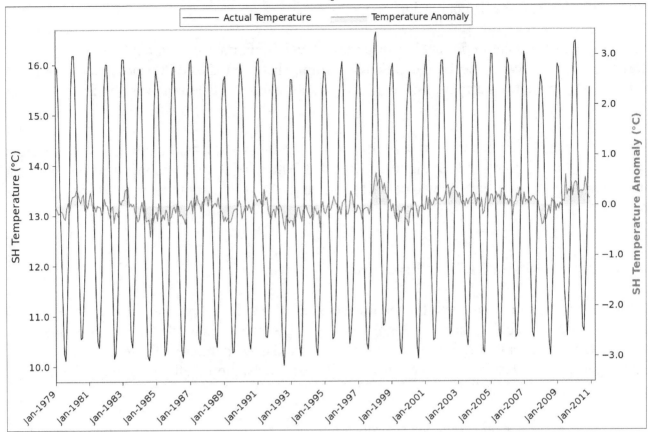

Illustration 97: The SH gets a +/- 3.5 °C range. The temperature anomaly is smaller for the SH, but as a percentage of the yearly change it is actually higher than the NH.

So while the NH has contributed to 67% of the warming in the past 10 years, greater variation of the NH temperature should be expected. No one should be surprised that the part of the Earth that sees the largest annual temperature swings also experience the largest temperature anomaly. Measuring the global temperature anomaly in this way will always ensure that the NH dominates the result.

Temperature Measurements

When I initially wrote this chapter I had not been heavily involved in the global warming debate. It was initially a much different chapter than it is now. Many of the articles on my website deal with aspects of the modern warming period. The debate about how temperatures are measured and what method is best are endless. Avoiding that debate in the chapter was difficult, but important.

The reason why is that it just doesn't matter. The Earth is on average a little warmer than it was 150 years ago. About the same difference that happens between any two weeks in the month of April. I see no reason to debate that. The purpose of this book is to cover the broad aspects and provide one source that will explain the Earth's climate and give the information a person needs to reach a reasonable conclusion.

The ocean cycles are important because they play an important part in explaining why there is short-term variability in the climate. The warming between the 1970 and 2000 is a perfect example of short-term warming that is mostly likely caused by natural variations in the ocean cycles.

Instead of this being the longest scientific content in the book, it will simply point to resources for the many, many topics that are involved in the debate with some basic information about each topic. That is really the only way to cover the amount of topics without turning this section into a book all by itself.

Station Measurement Method Problems:

I have written a little bit about this on the website. The simplest problem that exists is that this method shows a shocking lack of sensitivity to climate events that are known to happen. For example the ENSO has a big impact on the average global temperature. There is no question that this happens. In the last El Nino event from 2009-2010 the station methods failed to detect any of this change in temperature[118].

Some years it does and some years it does not, but one trend that is clear is that it detects the signal of the ENSO less and less each time it happens. Why this is happening is unclear, but the satellite methods do not suffer from this same lack of sensitivity.

The Urban Heat Island Effect (UHI) is another problem as most station data is collected in locations where people live. An airport might be the worst place on Earth to measure temperature used to monitor climate, but that is where many of the official measurement stations are located.

Steve McIntyre and Dr. Roy Spencer are two well known people who object to global warming. Mr. McIntyre is a Mathematician who has found numerous errors in the statistical study of temperature change. His blog http://climateaudit.org/ is a never-ending source of detail to the math of reconstructions and measuring the Earth's temperature. He will give you far more information than I could possibly give you and has even published peer-reviewed journal articles on the topic.

Dr. Roy Spencer[119] has a Ph. D. in meteorology and is a Senior Scientist for NASA in the study of Climate Studies. Note that he studies the climate and not climate change. He has a book titled *Climate Confusion* that also points out flaws in the idea of AGW and the problems associated with the politics of global warming. These two have both found numerous problems in how the station data has been used to reconstruct the modern temperature of the Earth.

The debate about the station method and the satellite method will go on until the global

166

warming debate ends.   Understanding the sides and the views is the best that can really be accomplished. The people who make the best predictions about the weather use the satellite data exclusively and that is enough for me. Predicting the weather 10 days in the future is difficult and it takes accurate satellite data to even attempt it. Ask a meteorologist and they will tell you, the satellite data is the way to go.

If the satellite data showed more warming, I would still use it because it is the most accurate and responsive to changes that take place. It is better at detecting cooling from volcanoes and warming from ENSO events. To me it is that simple. There is simply no reason to use an inferior method of measurement. The only reason the warmists don't like the satellite method is because it shows a lower rate of warming. Not liking the results is the worst scientific method for choosing a system of measurement.

Chapter 11   The Earth's Energy Balance

It is time to take a little breather from temperature charts. Hopefully there have been enough to show that the Earth does have a significant amount of temperature variability. There are the natural cycles that are always causing the weather around us to behave in unexpected ways. Climate is more predictable, but only when viewed in the proper long-term context. Knowing the average temperature for the month of March is not useful in planning for the coming weekend. The odds are generally low that any particular day will be average.

All of the information so far took a while for me to find. When I started trying to understand the topic of global warming it is the information in this chapter that I started with. What I quickly realized is that understanding the energy balance and the atmosphere is not enough to really answer the question of what to expect in the future. There is a reason why many people are convinced that the Theory of Global Warming is correct. That view is primarily based on the information that will be discussed in the next two chapters. Associating this information only with the fact that the Earth is warmer now than it was 100 years ago will lead to the conclusion that CO_2 levels have a powerful impact on the Earth's climate. I have no problem with people that take that view, but I tend to get agitated when they only regurgitate talking points without bothering to understand the science.

If I refer back to the Settlers that are trying to determine what climate to prepare for, there are only two real options that have been presented and there is no reason for any of them to suspect that things will stay the same. There are two contradictory theories that have been presented and they have to decide which path to follow. To refresh things a little bit, here is a quick summary.

Theory #1: The level of CO_2 had been increasing since June and in October
the temperature was rising. The theory presented was that
CO_2 levels controlled the temperature of the region they were settling.

Theory #2: The amount of sunlight each day was decreasing due to
the Earth's orbit. As the amount of sunlight continued to decrease, the temperature
would drop, even if the temperature trend for the past week was
increasing temperatures in mid-October.

There is one thing that both of these theories have in common. That common thread is energy. It takes energy to cause something to warm up. When something cools down it loses energy. In Theory #2, it is clear what the change in energy is. Less energy is coming from the sun so the temperature drops. That part is very clear.

For Theory #1 the issue of energy is more complicated. To understand Theory #1 requires an understanding of how energy leaves the Earth. That is because much like the Earth is always getting energy from the Sun, the Earth is also always losing energy to space. The Earth is at the temperature that it is because the energy arriving from the sun and leaving the Earth is balanced. Or as the warmists would say, it used to be balanced and CO_2 emissions have now upset the balance and will continue to upset the balance for as long as the amount of CO_2 in the atmosphere continues to be over 350 ppm.

So understanding the role that CO_2 has in the Earth's climate requires that the energy balance of the Earth be understood. The overall energy balance of the Earth is very simple. It is only when more and more detail is added that the complexity becomes apparent. In the simplest form the energy balance of the Earth looks like this.

Energy from the Sun = Energy lost to space

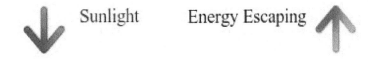

In a simple situation a change in the incoming energy will result in an almost immediate change in the energy escaping. The situation for the Earth is anything but simple, but the theory of global warming proposes that CO_2 in the atmosphere has a net effect of reducing the energy that is escaping.

Energy from the Sun (stays the same) = (decrease) Energy lost to space + warming

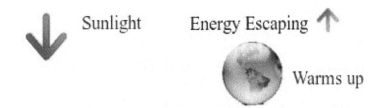

That is the basic idea behind the Theory of Global warming: that the increasing level of CO_2 in the atmosphere will cause the amount of energy leaving the Earth to decrease which will result in the Earth warming up[120].

Fortunately, the flow of energy from the sun is a very well-studied topic and has been for a long time. I will use the NASA Earth Observatory[121] information to explain the flow of energy from the sun, to the surface and then away from the Earth. In many ways it is very similar to the water cycle that most people learned in school.

Each flow of water here is much like each flow of energy. The source is the Sun and in the end the energy leaves the Earth. In the water cycle it might take a long time for the water to end up back in the ocean and sometimes it takes a very long time for the energy from the sun to leave the Earth. The question is; will CO_2 upset the cycle enough to cause a change to the overall energy cycle?

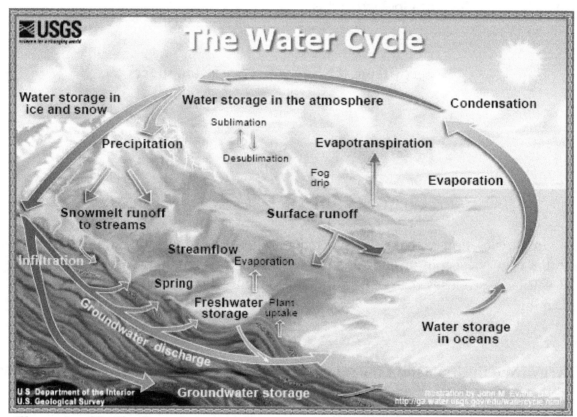

Illustration 98: The flow of water in the water cycle is similar to the flow of energy in the energy balance for the Earth.

In addition to the percentages that the NASA Earth Observatory uses, I will be using the actual numbers from the 2008 Kiehl-Trenberth (KT08) paper[122]. Since there are two sources involved, there can be slight differences between the two. In the science section there is a chart that shows other papers and each of them also arrive at slightly different values.

The units for energy in this chapter are W/m^2. That is watts of energy per meter squared. This is identical units to the insolation from the Sun. This means that the solar panel and light-bulb analogy from Chapter 6 applies to this as well. Energy of 50 W/m^2 is enough to power a 50W light-bulb. All energy units that I use are in those units. Just think of it in terms of light-bulbs

170

that a 1 meter squared solar panel will power.

Before diving into the details of the energy balance I will start with a broad overview. This is a variation of a commonly used energy balance diagram. The modifications I have made to it will be discussed later in the chapter and in depth in the Scientific Content.

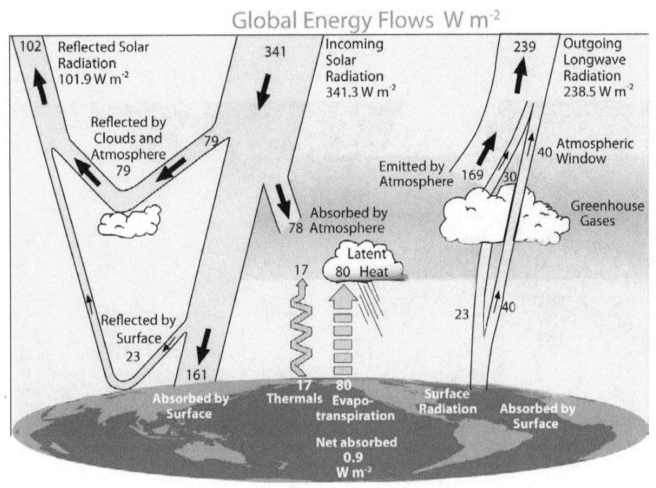

Illustration 99: This is a modified version of the 2008 Kiehl, Trenberth paper that discusses the Earth's energy budget. It has also been used in many other forums. It and the 1997 version (shown in the Science Content) are perhaps the most commonly used graphic for the Earth's energy balance. See the Science Content for a discussion of why it has been modified from it's original form.

The end result is that the energy leaving the Earth is equal to the energy coming from the Sun. It just shows the different paths that the energy takes before it leaves the Earth. This is the energy that drives almost everything that takes place on the Earth from gentle breezes to hurricanes and even the ocean currents.

Energy From the Sun:

This part is really simple. Pretty much all the energy available on the Earth comes from the sun. The sun sends out lots of energy in all directions and the tiny fraction that hits the earth is all that the Earth receives. The earth does have some energy deep within and volcanoes and deep sea vents are places where the Earth has energy that can leak out, but it is small enough that it can be ignored.

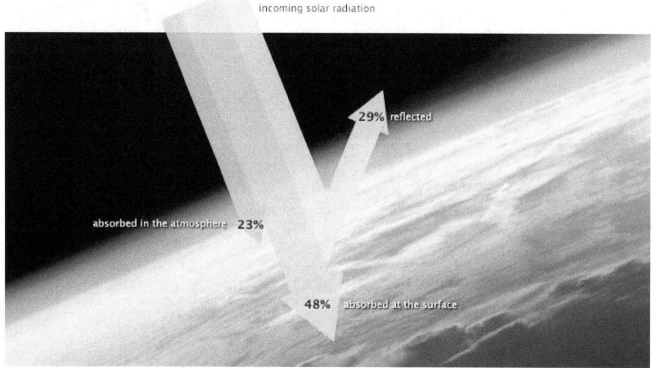

incoming solar radiation

29% reflected

absorbed in the atmosphere 23%

48% absorbed at the surface

Illustration 100: I am using the pictures from the NASA Earth Observatory. Due to rounding and such, different sources have slightly different values for each energy stream. The TSI% in the chart is the comparable values to this graphic.

Total Solar Insolation (TSI)			%	TSI %
Cloud Reflected	79	W/m²	23%	23%
Surface Reflected	23	W/m²	7%	7%
Earth Absorbed Energy (EAE)	161	W/m²	47%	47%
Atmosphere Absorbed	78	W/m²	23%	23%
Total TSI	341	W/m²	100%	100%
ASR	239.0	W/m²	70%	70%
Reflected	101.9	W/m²	30%	30%

Illustration 101: 30% of the Sun's energy is reflected away and has no impact on the Earth. The remaining 70% warms the Earth up.

Four things happen to the energy from the sun. It can be reflected away by clouds or the Earth's surface. ~30% of the total energy from the sun is reflected away is never absorbed by the Earth. This energy is just passing by and really doesn't do anything.

The part that matters is the Absorbed Solar Radiation (ASR). That is the energy that in some way gets absorbed by the Earth. The total amount that is absorbed is 239 W/m². This is the energy that is used by the Earth. Without that energy there would be little different between the Earth's temperature and that of Pluto.

The energy that is absorbed by the Earth is also approximately equal to the energy that is always leaving the Earth. That difference between incoming and outgoing is the total sum of energy that is used by the Earth. Photosynthesis in plants would be one way that the Earth traps energy here on Earth. Photosynthesis converts the sunlight into chemical energy that can be used by the plant (or animal, fungus or even car) in the future. Fossil fuels are really just reserves of very old solar energy that was converted into stored chemical energy a long time ago.

When volcanoes emit large quantities of material into the stratosphere, this reduces the energy that reaches the Earth's surface which in turn causes the Earth to cool down. The amount of energy that is reflected away increases and as a result there is a drop in the Earth's absorbed energy. The effect is noticeable cooling that takes place within a few months. Most of the delay is simply the time it takes for the material to spread around the Earth. When Mt. Pinatubo erupted on June 15th of 1991 it took about 6 months for the

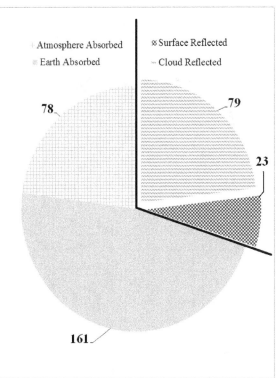

Illustration 102: The Earth Absorbed energy is the key part for keeping the Earth warm. It is that energy that provides the energy for the GHE.

173

material to cover the entire stratosphere. At that point the cooling effect of the eruption was global and the result was a global drop in temperature.

There are two ways in which the energy from the Sun is absorbed. The first part of the Earth to absorb the energy is the stratosphere which gets 78 W/m^2. This is where the ozone layer blocks the high energy UV rays from the sun. All of the UV that is blocked is absorbed by the stratosphere. This causes warming in the stratosphere and it also causes the ozone layer to be constantly regenerated. Ozone and oxygen both play a role in blocking the UV energy in the stratosphere.   When oxygen absorbs the UV the process is started to create ozone. Most of the energy that is absorbed by the atmosphere is absorbed in the Earth's stratosphere. Some does get absorbed by water vapor in the upper troposphere, but it is a very small amount in comparison to the effect of the stratosphere.

The remainder of the ASR is the energy that is absorbed by the land and oceans. This 161 W/m^2 is what provides the warmth and energy for the Earth. The oceans are liquid because of that energy and plants grow as a result of that energy. Everything from wind to ocean currents is directly powered by the energy that is absorbed by the surface of the Earth.

All of the energy that is absorbed at the Earth's surface is altered into another form of energy. In a sense all the surface of the Earth does is convert insolation from the sun into the heat energy that the Earth can use. The heat then leaves the Earth's surface in a variety of ways which result in the atmosphere warming up. This is the fundamental cause of the Greenhouse Effect (GHE).

Surface Absorbed Energy and the Greenhouse Effect:

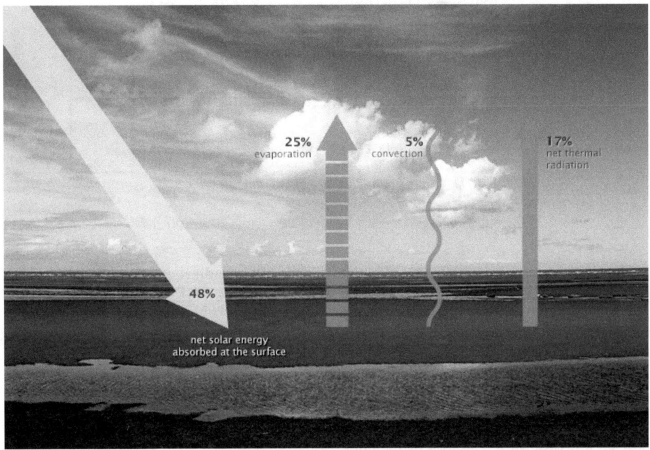

Illustration 103: Convection and Thermals are commonly used by different sources. The TSI% in the chart is the comparable values to this graphic.

Once the sunlight is converted to heat it behaves in a way that people are familiar with. Think of a pot of boiling water. If you take it off the stove it will stop boiling and start to cool down right away. That is exactly what is always happening at the Earth's surface. The heat is moving from the warmest place (surface) to the cooler air above. These are three ways that energy leaves the Earth's surface: evaporation, convection (or thermals) and as infra-red energy (thermal radiation).

The ~120 W/m² of energy that then flows to the atmosphere is what causes the Earth's GHE. It is the Greenhouse Effect that keeps the Earth warm. Without the atmosphere the Earth would be significantly colder than it is now. Not all of the energy that is absorbed by the Earth's surface is transferred to the atmosphere. There are two parts of what is labeled in the above picture as *Net Thermal Radiation*. I have them split out as Net Thermal Radiation (GHE) and (Window) depending on where the energy goes.

Earth Absorbed Energy (EAE)			%	TSI %
Convection (Thermals)	17.0	W/m²	11%	5%
Evaporation	80.0	W/m²	50%	23%
Net Thermal Radiation (RHT)	23.0	W/m²	14%	7%
Total to atmosphere (GHE)	120.0	W/m²	75%	35%
Net Thermal Radiation (Window)	40.0	W/m²	25%	12%
Used by Earth	0.9	W/m²	1%	0.3%
Total EAE	160.9	W/m²	100%	47%

Illustration 104: The energy that transfers to the atmosphere is what causes the Greenhouse Effect. The 'window' is infra-red energy that is NOT absorbed by the atmosphere. See Scientific Content for details on the 'window'

The Greenhouse Effect (GHE) states that the surface of the Earth is warmer than it would be if there was no atmosphere. The most common estimate[123] of the GHE is that the Earth is 33 °C warmer as a result of the atmosphere. There is a much more extensive discussion of the GHE in the Science Content of this chapter and the next.

What is important now is the 120 W/m² of energy transferred from the surface and the 33 °C total effect of the GHE. Those two numbers will show up a great deal for the rest of the book. I will also break them down in several different ways. The 120 W/m² was the first clear indicator on my path to reaching the conclusion that global warming is not a threat. Some people will consider it controversial, but it is the correct value for the amount of energy that is transferred to the Earth's atmosphere.

Evaporation is the dominant factor in energy being transferred from the surface of the Earth to the atmosphere and as such plays the biggest role in the GHE. At 80 W/m², the transfer of water vapor from the surface through evaporation to the atmosphere where it condenses accounts for 67% of all the energy that the atmosphere gets from the surface. This is also called latent heat transfer.

This method of energy transfer is what causes warm fronts of moist air that travel from warm

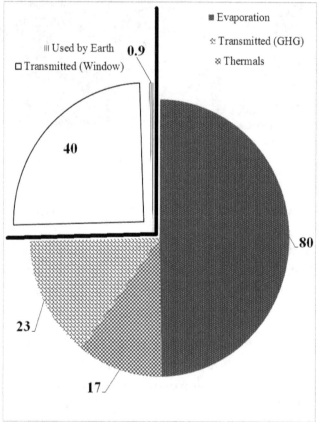

Illustration 105: 75% of the energy that is absorbed by the surface is then transferred to the atmosphere. Only the 25% in the infrared 'window' escapes directly to space.

open water to other locations. The Gulf of Mexico is a perfect example. The warm surface waters in the Gulf cause high amounts of water to evaporate. This warm, moist air then travels northward into the Midwest portions of the United States where it warms the cooler, drier air that is flowing westward. The Midwest is warmer and has more humidity as a result of this type of energy transfer. Even lake effect snow is an example of this type of energy transfer. In all cases the water vapor carries energy from liquid water on the surface and transfers energy to the atmosphere when it condenses.

Since water covers 71% of the Earth's surface it should be no surprise that it is responsible for most of the energy that is transferred to the atmosphere. Since evaporation is the primary method of energy transfer by water, evaporation would be expected to be the dominant method of energy transfer to the atmosphere. Once again geography plays an important role in defining how the Earth responds to energy from the Sun. If the surface area of the Earth had less water and more land, the role that evaporation plays would be smaller.

Convection is a natural circulation of air that happens as a result of differing temperatures of the air. Warm air rises, so when cool air is in contact with the warm surface, the air warms up and rises. This in turn causes cool air to drop to the surface. In the energy balance this natural convection has been labeled as Thermals. This energy flow is the smallest at 17 W/m². This means that thermals are the least important contributor to the Greenhouse Effect at 14%.

Thermals are interesting because the temperature difference between the surface and the atmosphere determine how much energy is transferred. If there is a small temperature difference, then there is very little convection. If there is a large difference, there is a large amount of convection. Places like deserts and many man-made things like roads and buildings can generate high surface temperatures that drive higher rates of convection. Convection of humid air is also a key factor in generating afternoon thunderstorms.

The warm/moist air rises into the cool dry air above it. This causes the water to condense out which provides energy to the air and also changes the density of the air. This combination of convection and evaporation/condensation create powerful thunderstorms in places where the right mixture of these conditions exists. It is these thunderstorms that also generate most tornadoes. The cool dry air above is every bit as important as the humid air near the surface. That is why tornadoes happen more often in what is called Tornado Alley in the United States. That is the most active location on Earth for tornadoes precisely because of the combination of humid air near the surface and cool dry air at higher altitudes.

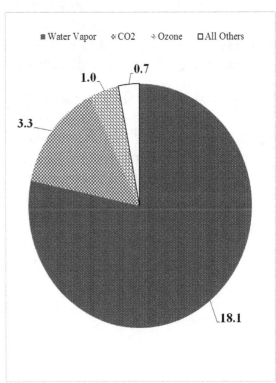

Illustration 106: Final contribution of greenhouse gases to the total 23 W/m2 of RHT to the atmosphere.

177

Together Evaporation (~67%) and Thermals (~14%) account for 81% of all the energy that is transferred from the surface to the atmosphere. In neither case does the amount of CO_2 in the atmosphere alter them (although it has been proposed as a feedback). If the Earth's atmosphere had no CO_2 at all, the amount of energy transferred to the atmosphere by these methods would be unchanged. That leaves only ~19% of the energy that could be affected by the amount of CO_2 in the atmosphere.

The NASA Earth Observatory (Illustrations 100, 103, 110) calls this transmitted energy Net Thermal Radiation. Engineers call this Radiative Heat Transfer or RHT. There are two parts to the RHT from the Earth's surface. There is the part that goes to the atmosphere and there is the part that goes straight to space. The amount of energy that leaves the surface of the Earth is 63 W/m2, but of that only 23 W/m2. goes to the atmosphere. The larger portion (40 W/m2) travels straight through the atmosphere and into space. It is that portion that is designated by the infrared window[3].

In order to simplify, I will treat the smaller flow of energy that is absorbed by the atmosphere and the larger flow that escapes through the infrared window as different flows of energy, even though they come from the same source. The atmospheric absorbed portion accounts for 23 W/m^2 and is slightly larger than the amount of energy transferred by convection which means it accounts for 19% of the energy that the atmosphere absorbs from the surface.

One confusing aspect of the global warming debate is that radiative heat transfer is also sometimes described as the Greenhouse Effect by itself. That further complicates discussions about the energy balance. In addition, each gas that is called a greenhouse gas has a different effect. This makes the task of precision difficult, but one reason why I wrote this book was to try to make it understandable. I will start with the most dominant of the greenhouse gases. By itself this one gas is more important than all the others combined.

That most important greenhouse gas is water vapor. It is enormously more important than CO_2 as a greenhouse gas. In fact, it dominates the RHT portion of the energy flow even though it is extremely variable[124] in the atmosphere and plays a critical role in cloud formation. The difference in RHT from the surface to the atmosphere when there are clouds in comparison to not having clouds is large. If that wasn't enough, different clouds also affect RHT in different ways.

There is a more detailed discussion in the Scientific Content[125], but the final result for the different greenhouses gases is:

---

3   See Science Content in the next chapter for in-depth discussion about the infrared window.

62% Cloudiness			%	TSI %
Water Vapor	18.1	W/m²	79%	5.3%
Carbon Dioxide	3.3	W/m²	14%	1.0%
Ozone	1.0	W/m²	4%	0.3%
CH4 + N2O	0.7	W/m²	3%	0.2%
Total 62% Cloudiness	**23.0**	**W/m²**	**100%**	**6.7%**

Illustration 107: Energy contribution of each greenhouse gas to the total energy transferred to the atmosphere by radiative heat transfer.

So not only is water vapor the key contributor to the GHE by nature of evaporation, it is also the most important greenhouse gas by a large margin. CO_2 is not insignificant in it's contribution to this portion of energy transfer, but at 3.3 W/m2 it accounts for only 1.0 % of the total energy balance. When only greenhouse gases are considered it accounts for 14% of the energy.

This is where terminology leads to confusion about the Earth's climate. It is total energy transfer that matters to the Earth's climate. Oftentimes people only discuss the heat transfer by the greenhouse gases. In that discussion CO_2 will have a more prominent role than it does in the total energy balance. The combined effect of all greenhouse gases on the GHE is only 19%. CO_2 is only critical in a discussion of that 19% of the total energy balance.

There is nothing special about the energy absorbed by greenhouse gases when compared to the other methods of transferring energy from the surface to the atmosphere. There will be more about this in the next chapter which focuses more on the atmosphere. Since this chapter discusses the energy balance, I will try to stay on topic.

These are all of the flows of energy that leave the Earth's surface. Since it is the 120 W/m² that are so critical, I would like to show them all together by size of their individual contribution. Water vapor by evaporation and as a greenhouse gas accounts for 82% (67% and 15%). This means that water vapor is responsible for 82% of the Earth's Greenhouse Effect. The Earth truly is a water planet.

The gas that garners so much attention is 4th on the list. It accounts for 2.8% of the total Greenhouse Effect. While not insignificant, it is far from critical to the whole. If the entire 120 W/m² is needed for the Earth to be 33 °C warmer, then CO_2 is responsible for 2.8% of that 33 °C. That by itself is one of the real problems for the theory of global warming.

Greenhouse Effect: Energy Sources			GHE %	TSI %
Evaporation	80.0	W/m²	66.7%	23.5%
Water Vapor (GHG)	18.1	W/m²	15.1%	5.3%
Convection	17.0	W/m²	14.2%	5.0%
CO2 (GHG)	3.3	W/m²	2.8%	1.0%
Ozone (GHG)	1.0	W/m²	0.8%	0.3%
Other GHG	0.7	W/m²	0.6%	0.2%
Total GHE	**120.0**	**W/m²**	**100%**	**35%**

Illustration 108: This is a list of the energy flows that cause the Greenhouse Effect that keeps the Earth warm.

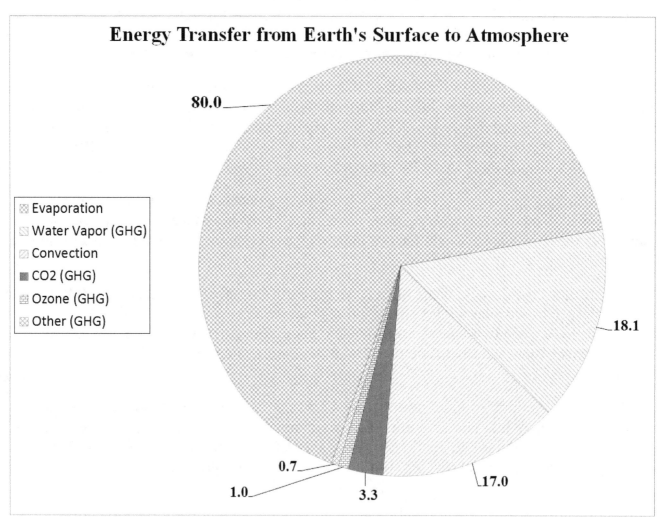

Illustration 109: The 6 main components of the Earth's Greenhouse Effect. The total energy for them is the 120 W/m². GHG is energy absorbed by specific greenhouse gases.

Atmospheric Energy Balance:

Almost all the energy from the Sun ends up in the atmosphere at some point. The three flows of energy from the surface combine to form the single largest flow. All of the energy in the atmosphere leaves the Earth as infra-red energy. The energy leaving the Earth is almost equal to the energy that the Earth receives from the Sun.

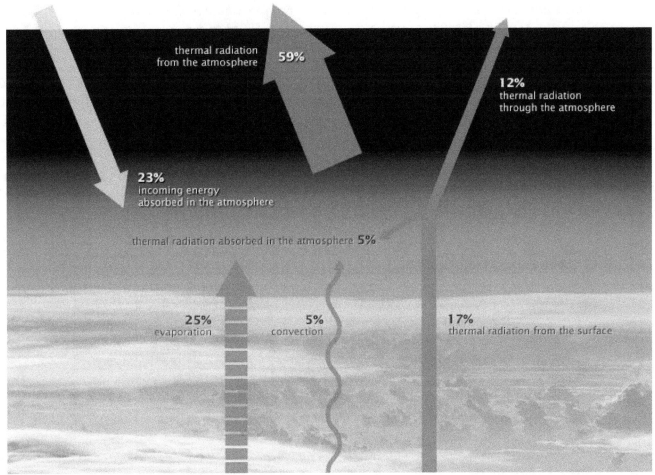

Illustration 110: Only 5% of the total energy is absorbed by greenhouse gases. Most of the thermal radiation from the surface escapes directly into space. The atmosphere as a whole is ALWAYS losing energy to space. Without the constant flow of energy upwards from the surface, the atmosphere would quickly cool down. The TSI% in the chart is the comparable values to this graphic.

Atmosphere Absorbed (AA)		%	TSI %
SW Absorbed (Stratosphere)	78.0 W/m²	39%	23%
Convection (Thermals)	17.0 W/m²	9%	5%
Evaporation	80.0 W/m²	40%	23%
Net Thermal Radiation (RHT)	23.0 W/m²	12%	7%
Total AA	**198.0 W/m²**	**100%**	**58%**

Illustration 111: All the energy flows from the surface combine in the atmosphere and leave the Earth's atmosphere by RHT to space.

An example of energy that is used by the Earth is plants. They convert the Sun's energy into chemical energy through photosynthesis. That chemical energy then enters the food-chain where it is used in a variety of ways. Fossil fuels are also an example of that type of energy that was converted long ago by plants.

Now that each flow of energy has been discussed in detail it is possible to show the overall energy balance for the Earth. The most critical 120 W/m² does not show up in the overall balance because it is mixed in with the other flows.

Total Solar Insolation (TSI)		%	TSI %
Cloud Reflected	79 W/m²	23%	23%
Surface Reflected	23 W/m²	7%	7%
Earth Absorbed Energy (EAE)	161 W/m²	47%	47%
Atmosphere Absorbed	78 W/m²	23%	23%
Total TSI	**341 W/m²**	**100%**	**100%**
ASR	239.0 W/m²	70%	70%
Reflected	101.9 W/m²	30%	30%

Outgoing Energy (OE)		%	TSI %
Cloud Reflected (SW)	79.0 W/m²	23%	23%
Surface Reflected (SW)	23.0 W/m²	7%	7%
Window (LW)	40.0 W/m²	12%	12%
Atmosphere Emitted (LW)	198.0 W/m²	58%	58%
Total OE	**340.0 W/m²**	**100%**	**100%**

Illustration 112: This table sums up the flows of energy from the Sun and shows how they all leave the Earth. The pictures from the NASA Earth observatory use percentages of the Sun's incoming energy which is the 341 W/m².

182

The full energy balance for the Earth is tabulated in Illustration 111. Other papers will get slightly different numbers for each of the energy flows and that is to be expected. This is a good annual estimate for the total energy balance. Of the entire energy balance, CO_2 is involved in only ~1%.

It must be understood that this is the annual average. The energy balance in July and January are different from this. The difference in the energy balance between those two months will serve to prove just how important the Northern Hemisphere is in determining the Earth's temperature. Purely as a side effect it will further weaken the theory of global warming.

Earlier I mentioned that the Earth is closest to the Sun in early January[126]. It is at that point of the year that the Earth gets the most energy from the Sun. The total energy at that point in year is 353 W/m2. The Earth is the coldest when it is getting the most energy from the Sun. That is because the Southern Hemisphere is getting that energy instead of the Northern Hemisphere.

When the Earth is the warmest in July, it is getting only 329 W/m^2 of energy from the Sun. The Earth is 4 °C warmer when it is getting 24 W/m^2 less energy from the Sun. That is a 7% difference in the total energy that the Earth is getting, but the Earth's temperature is opposite of the total energy it gets from the Sun. The Earth cools down while the energy is increasing and it warms up while the energy is decreasing.

Why is the Earth 4°C cooler when it is getting 7% more energy from the Sun? The answer is the difference in geography between the Southern and Northern Hemispheres. The land in the NH warms up more quickly with less energy. It also cools down more quickly when the energy is not present. The SH only has a 5.5 °C difference between summer and winter. The NH has a difference of 13 °C. It gets warmer and colder than the SH in each of the respective seasons. So even though the SH has Antarctica, the NH gets colder during the winter because more of the NH has a large drop in temperature.

Solar Energy and Temperature of Earth

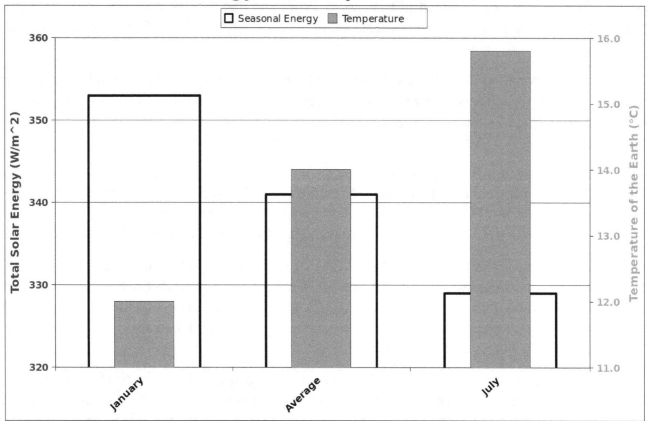

Illustration 113: The Earth behaves opposite of the Total Solar Insolation because the different hemispheres respond differently. In effect the Earth has two separate energy balances and oscillates between the two of them.

The main warmist argument against the Milankovitch Cycle is that there is not enough of a change in Total Solar Insolation (TSI) to explain the changes in global temperature. Meanwhile they ignore the fact that each yearly cycle proves that TSI does not drive global temperatures as much as regional insolation does. The Earth is warmest with the lowest total energy, because that is when the NH is in summer.

This is one case where it is reasonable to compare the Earth to the Moon. The Moon shows the expected temperature variation between January and July. The Moon is 6 °C warmer in January than it is in July. That is because it responds more directly to the amount of total energy from the Sun. That the Earth has the opposite response to the total amount of energy from the Sun demonstrates the importance that the differing geography of each hemisphere plays in controlling the Earth's climate. The Moon follows the TSI theory, but the Earth does not.

The geography of the Earth makes more difference in temperature than a 24 W/m² change in total energy from the Sun. This is why the Milankovitch Cycle is capable of making such a difference in the Earth's climate. The regional changes in solar insolation are the driving factor in the climate cycles. The change in the Earth's temperature follows the hemispheric insolation that the Northern Hemisphere receives.

184

Scale of Energy and Temperature Changes

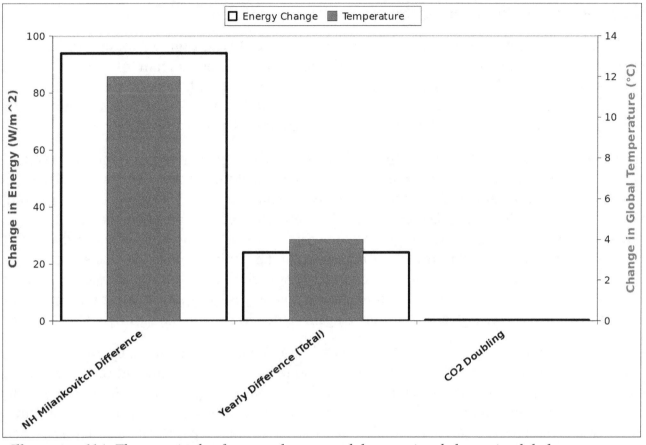

Illustration 114: The magnitude of energy changes and the associated change in global temperature.

The total range in the 65N summer insolation is 94 W/m2 over the past climate cycle. Considering how strongly the Earth's overall temperature responds to the seasons of the NH, that magnitude of energy change provides a clear mechanism to cause the climate cycle. There is no reason to doubt that the Milankovitch Cycle is the driving force in the glacial-interglacial cycle.

Warmists do have a counter theory. They argue that any increase of CO_2 in the atmosphere will increase the warming effect of the greenhouse gas processes[127]. In essence each doubling of the CO_2 concentration in the atmosphere will cause a ~10% increase in the total effect of CO_2. Since the total energy effect of CO_2 is 3.3 W/m^2, a 10% increase would cause a total increase the total effect of CO_2 to a total of 3.6 W/m^2. Obviously that makes little sense, but they do phrase it in a different manner. The major difference is that they also claim that the entire Greenhouse Effect is caused almost completely by the greenhouse gases and the other energy flows (evaporation, thermals) are secondary to the greenhouse gas effect. That is what the next chapter will deal with.

So for now I will proceed with the energy balance that has been proposed. I fully accept that significant changes to the energy balance will cause significant changes to the Earth's temperature. The Milankovitch Cycles cause the energy balance to change in the NH which results in either excess energy or an energy deficiency, the former causes warming and the later causes cooling. 16,000 years ago the NH started to get excess energy and it warmed up. For the past 3,000 years the NH has been running an energy deficit.

The real question facing mankind now isn't will CO_2 cause global warming, but is CO_2 capable of preventing the next glacial? When I compare the amount of energy that CO_2 contributes to the Earth's atmosphere it does not look promising. The annual TSI variation is ~ 7 times larger than the total contribution of CO_2 to the GHE. A 10% change in a small overall contributor is not going to make any measurable difference.

The current state of the Earth's energy is that the SH is getting positive summer insolation because it is the hemisphere that is pointing to the Sun when the Earth is closest to the Sun. For that reason alone it is reasonable to expect it to be warming up. The SH has been getting positive summer energy for the past 6,000 years. The Antarctic ice cores have shown cooling for the past 6,000 years while getting more energy from the Sun. That is a powerful indicator that the NH summer energy is currently the key factor in regulating the Earth's climate. In that same period of time the NH has lost 32 W/m^2 of summer energy. That is a 7% drop in summer energy. The only possible result of such a decrease in summer energy is a gradual decrease in summer temperatures. That is why glaciers have once again started to form in the Northern Hemisphere.

With this in mind, let's refer back to Theory #1 and Theory #2 that are really about the energy balance of the Earth. The difference between those theories is which amount of energy is more important. Theory #2 focuses on the amount of energy that the region is getting from the Sun to predict what the future climate will be. Theory #1 focuses on the idea that increasing the amount of CO_2 in the atmosphere will result in the atmosphere warming up because less and less energy will escape the Earth as a result of increasing CO_2 levels in the atmosphere. Both theories are correct in their way, but the difference is the size of energy involved.

There is one central difference in how the two theories account for the method of energy transferred from the surface to the atmosphere by radiative heat transfer. The differences will be presented in the Scientific Content of this chapter. The whole debate really does center around energy and how changing the concentration of CO_2 in the atmosphere will change the energy balance of the Earth.

If CO_2 does not cause a significant change in the Earth's energy balance, then the theory of global warming is not possible. The natural state of the Earth today is the long-term cooling trend as the interglacial season ends. This is based on how the Earth has responded to the energy from the Sun for the past million years. If CO_2 levels were stable, there would be no scientific basis to argue against that point.

The theory of global warming proposes a theory that the increasing levels of CO_2 are caused by mankind and that those increased CO_2 levels will disrupt the entire cycle that has been happening for more than a million years. In that period of time the level of CO_2 has varied a great deal, but at no time did an elevated CO_2 level prevent a glacial from starting.

I personally believe that the interglacial state of the Earth's climate is highly preferable to the glacial state. The Earth has unquestionably been moving towards the glacial state and the question I have to ask is, can CO_2 stop the next glacial? Based on the contribution of CO_2 to the energy balance it would appear improbable.

Warning:

Scientific Content!!!

An Engineering Approach to the Energy Balance:

If a pot of water that is 50 °C with a sealed lid is placed in an adiabatic, sealed room that is 20 °C, what will happen? The simple answer is that energy will leave the pot of water and the pot of water will cool down. The energy will transfer to the air in the room until they are at the same temperature. When they are the same temperature, the net energy transfer between the pot of water and the air in the room will be zero. That is not to say that there is no energy transfer taking place, but there is no longer a net transfer of energy between the pot of water and the air in the room. When the net energy transfer is zero, there is no change in temperature.

At that point the temperature of the room and the pot will be the same, which will be between 20 °C and 50 °C. The larger the pot is in comparison to the room, the closer to 50 °C the final temperature will be. This is a basic example of the conservation of energy. Energy cannot be created in either case, but it will move from one place to another.

In this example there are only two methods that energy will transfer from the pot to the room. There will be convection and radiative heat transfer. That is because the pot is sealed and water vapor cannot escape the pot to warm up the room. If the pot were open, the evaporation of water would speed up the process of reaching the new equilibrium temperature. Using only convection and RHT the process will be slower. This is especially true because the rate of heat transfer for convection and RHT is dependent on the temperature difference between the two objects.

The first step to understanding radiative heat transfer is to understand that all objects that have a temperature radiate energy. The amount of energy radiated by an object is:

$$Q = \sigma T^4$$

Where σ is the Stefan-Boltzmann constant of 5.67E-8 W/m2K4. The temperature is in absolute Kelvins. The energy radiated away from a blackbody object is strictly determined by the object's temperature.

The amount of energy transferred between two blackbody objects is simply the difference between the energy fluxes of each object. It is commonly described as:

$$Q = \sigma(T_2^4 - T_1^4)$$

So for the case stated above the energy transfer from the 50 °C (323K) to the 20 °C (293K) room is:

$$Q = 5.67E\text{-}8 * (323^4 - 293^4)$$

$$Q = 200 \ W$$

By definition, net energy can ONLY flow from the warmer body to the cooler. So in this situation the pot will lose 200 watts of energy by RHT to the room. If the pot were 25 °C (298K) instead, the heat transfer would only be 29W. Here is the energy transfer from the pot to the room for the whole temperature scale from 0 to 100 °C.

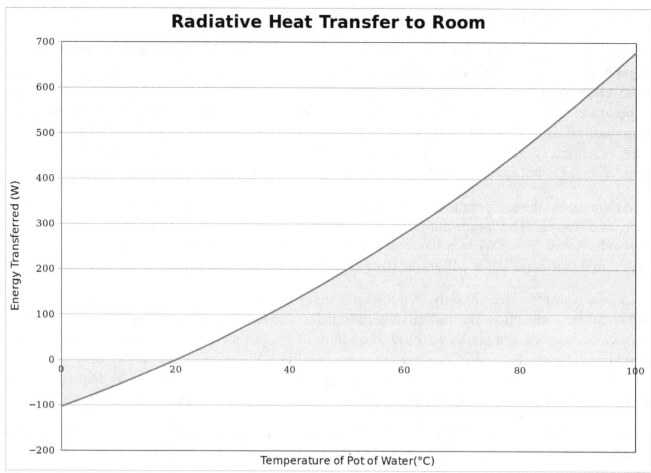

Illustration 115: Fixed room temperature of 20 °C. The greater the temperature difference the more energy is transferred.

The meaning of negative energy transfer is simply the change of direction. When the pot is colder than the room, energy is transferred from the room to the pot of water by RHT. The overall flow of energy always naturally occurs from warmer to cooler.

This is the basis for all radiative heat transfer between objects. This applies to what is called blackbody radiation, which means that the bodies absorb and emit energy perfectly. In practice the Earth's surface and the atmosphere are reasonably close to this. Ocean water for example is ~99% a blackbody, so the error is only 1% for ocean water. Overall the error for the blackbody assumption doesn't exceed 5% so it is a reasonable assumption to treat the Earth's surface as a blackbody. The atmosphere is more complex, but the same principle applies.

The transfer of energy is primarily dependent on the temperature of the objects if they are approximately blackbody objects. For this section that focuses on the net energy transfer I will proceed on the assumption that the Earth is a blackbody and the atmosphere acts in two ways. Part of it is a blackbody and part of it lets energy pass through without any absorption.

Since the normally measured temperature of the Earth is actually a measurement of the air 1.5 m above the Earth's surface, an estimate must be made for the difference between the air and ground. The temperature of the actual ground of the Earth is estimated to be ~2.5 °C[128] higher than the air 1.5 m above the surface. So instead of using 14 °C, I will use 16.5 °C for the temperature of the actual surface of the Earth. The radiated energy from a blackbody object at 16.5 °C is 399 W/m^2. Since the Earth loses 63 W/m^2 of energy from the surface by RHT, the effective energy flux of the atmosphere is 336 W/m^2 which corresponds to an absolute temperature of 277.5K (4.3 °C). That in turn corresponds to an altitude of 2.5km.

This is certainly a reasonable result for the net transfer altitude of energy from the surface to the atmosphere as will be explained in the next chapter. In effect the Earth's surface is transferring 63 W/m^2 of energy to an altitude of 2.2km. 23 W/m^2 of that energy is absorbed by the atmosphere and the other 40 W/m^2 escapes the atmosphere and into space.

That is the net flow of radiative energy from the surface to the atmosphere. I would also add one additional step and determine the effective temperature at which the Earth radiates energy to space. That effective radiative flux of the atmosphere is 198 W/m^2. That corresponds to a temperature of 243K (-30 °C). That is the temperature of the atmosphere at 9.4km (think 31,000 ft of altitude) which is the upper troposphere where the amount of water vapor in the atmosphere starts to really drop off. Since water vapor is the primary gas transmitting energy to space, the result is once again very reasonable.

Common Sense

In an everyday sense, consider a warm front that moves in after it has been cold. In that case the warm air warms up the surface of the Earth. When such an event takes place when there is snow on the ground, the snow starts to melt as a result of the warm air. In that case the atmosphere is warming the surface.

The reverse happens when a cold front moves into an area. The warm ground is then transferring energy to the air above it. Both situations do happen (such events also happen during the day/night cycle) on a regular basis. On *average*, the surface causes the atmosphere above it to warm because more energy is transferred from the surface to the atmosphere.

How the Energy Balance is Commonly Presented:

From an energy balance perspective this is very straightforward and there is little to argue about. Unfortunately this approach is not how such energy balances are typically done. Most of it is the same, but when it comes to the RHT portion of the energy balance, the typical approach of climate scientists is very different, even though the result is identical, they will argue that it is not.

Here is the energy balance presented by Trenberth and Kiehl in 1997[129] and is the most commonly used version of the energy balance. I used their updated 2008 paper for the energy balance I presented, so there are some slight differences.

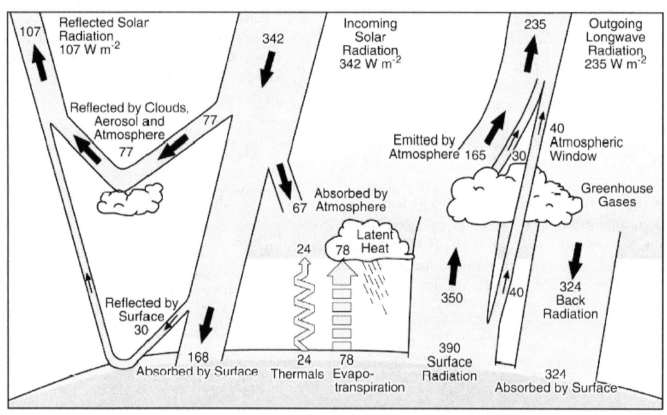

Illustration 116: Kiehl, Tranberth 1997 Energy Balance. Also used by the IPCC.

In effect there is no real difference between the energy balance I presented and the one displayed here. The appearance is completely different with regard to the RHT portion though. Instead of displaying the net energy flow from the Earth to the atmosphere, they display the energy flux from the surface to the atmosphere. That is an incredibly misleading decision. It gives the appearance that the surface transfers 230% more energy to the atmosphere than the sun transfers to the Earth's surface. Nothing could be further from the truth.

A consequence of that decision is that the energy flux from the atmosphere to the surface must be added in. Note that the energy flux from the atmosphere to the surface is 192% of the energy

192

the surface gets from the Sun. According to this energy balance, the atmosphere provides almost TWICE the energy to the surface that the Sun does.

As an engineer I find this energy balance to be intentionally misleading. In every other instance only the NET energy transfer is shown, but in the single aspect of RHT the energy fluxes are shown in lieu of the NET energy transfer. That back radiation of 324 W/m^2 is the single greatest evidence of the intent to mislead people and help ensure that people do not understand the Earth's climate.

While there is an energy flux of that magnitude, it is not an energy transfer in the accepted sense. If I refer back to the example of the pot of water in a room, the temperature of the pot only changes based on the net transfer of energy. I will alter the example slightly to prove this point.

I now heat the sealed pot up to 100 °C and then place the pot into an oven that is exactly 100 °C. The net energy transfer is exactly zero. The temperature will stay exactly the same as a result. In parlance of the Trenberth energy balance, the pot is losing 1097 W/m^2 to the oven. In this case the "back radiation" happens to be 1097 W/m^2 as well, but clearly the result is zero net energy transfer. Showing zero energy transfer by substituting in a 1097 W/m^2 energy flow from the pot to the oven and then adding in a 1097 W/m^2 flow from the oven to the pot isn't really a lie, but it is very misleading and it is clearly intentional.

Here is what the same energy balance would look like without the silliness added in by Trenberth. This energy balance now matches the percentages provided by NASA. While they can argue that the original is correct, this is the one that portrays the actual flow of energy.

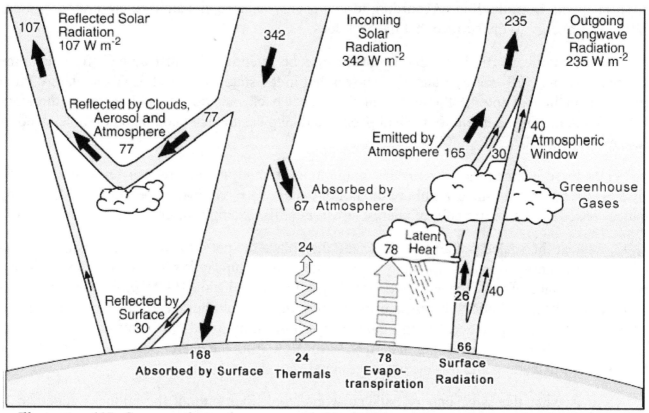

Illustration 117: Compare this to the one provided by NASA and the result is nearly identical. Both Keihl-Trenberth (1997, 2008) energy balances substitute radiative flux for net heat transfer. The numbers here are based on the 1997 paper.

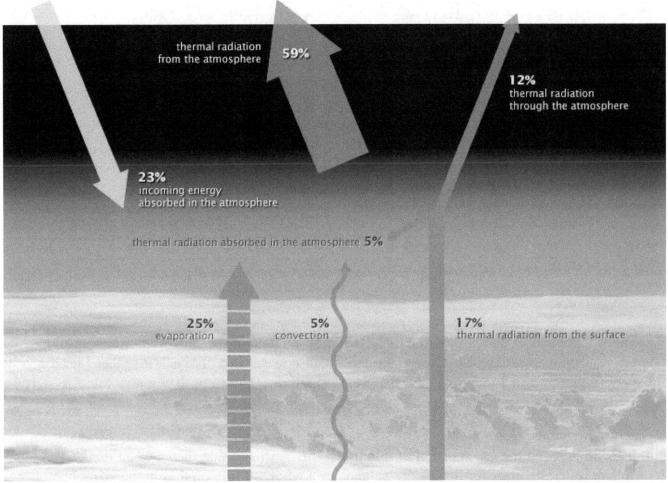

Illustration 118: NASA shows the net heat transfer only. The percentages here align almost exactly with the numbers I extracted from the KT08 paper.

The only explanation is that KT08 intentionally substituted radiative fluxes for heat transfers in order to make the greenhouse gas effect appear larger than it really is. It is a subtle and intentionally misleading diagram of the Earth's energy balance. That single misdirection has caused an enormous amount of misunderstanding about the Earth's energy balance.

It can also be demonstrated that it was intentional. In the KT08 paper, Table 2b shows the Net LW energy from the surface at 63 W/m². So he was fully aware of the correct value when he put the diagram together. That paper had the correct information, but they instead put in a misleading diagram that creates the impression that greenhouse gases play a larger role in the Earth's energy balance than the science shows.

TABLE 2b. Surface components of the annual mean energy budget for the globe, global land, and global ocean, except for atmospheric solar radiation absorbed (Solar absorb, left column), for the CERES period of Mar 2000 to May 2004 (W m^{-2}). Included are the solar absorbed at the surface (Solar down), reflected solar at the surface (Solar reflected), surface latent heat from evaporation (LH evaporation), sensible heat (SH), LW radiation up at the surface (Radiation up), LW downward radiation to the surface (Back radiation), net LW (Net LW), and net energy absorbed at the surface (NET down). HOAPS version 3 covers 80°S–80°N and is for 1988 to 2005. The values are from ISCCP-FD, NRA, JRA, and this paper. For the ocean, the ISCCP-FD is combined with HOAPS to provide a NET value.

Global	Solar absorbed	Net solar	Solar reflected	LH evaporation	SH	Radiation up	Back radiation	Net LW	NET down
ISCCP-FD	70.8	165.7	22.8	-	-	393.9	345.4	48.5	-
NRA	64.4	160.4	45.2	83.1	15.6	396.9	336.5	60.4	1.3
JRA	74.7	169.8	25.6	90.2	19.4	396.9	324.1	72.8	−12.6
This paper	78.2	161.2	23.1	80.0	17	396	333	63	0.9

Illustration 119: Buried in a table the KT08 paper shows the correct value for the LW RHT from the surface, but they failed to show that value in the main diagram.

The radiative flux that is described as 'Radiation up' and 'Back radiation' are simply the two parts of a normal radiative heat transfer situation that is described above. Note that I used nearly the same values when I demonstrated the RHT equation. The only difference was the starting surface temperature I used was slightly higher.

The greenhouse gases in the Earth's atmosphere only absorb 23 W/m^2 of energy from the surface. Changes to the CO_2 portion of that are so insignificant that they can be ignored as more trivial than rounding error for the energy balance as a whole. More on that in the next chapter.

When the actual diagram (Illustration 119) from the KT08 paper is shown in the unmodified form, the methodology to make greenhouse gases appear more important is evident. Instead of showing the actual amount of energy that is transferred (Net LW), the paper used the values 'Radiation Up' and 'Back radiation' from the table.

Does it make sense that the Earth's surface receives 207% more energy by that method than from the Sun? Especially considering that the atmosphere is colder than the Earth's surface? In all forms of heat transfer, energy flows from hot to cold, but this energy balance shows the Earth's surface getting twice as much energy from the atmosphere as it gets from the Sun.

That is exactly why in real-world applications of radiative heat transfer, it is only the NET transfer that matters because that is the only energy that can cause a change in temperature. Two objects that have a NET transfer of zero will not change temperature. That is what makes the energy balance below the single most misleading graphic in the entire global warming debate.

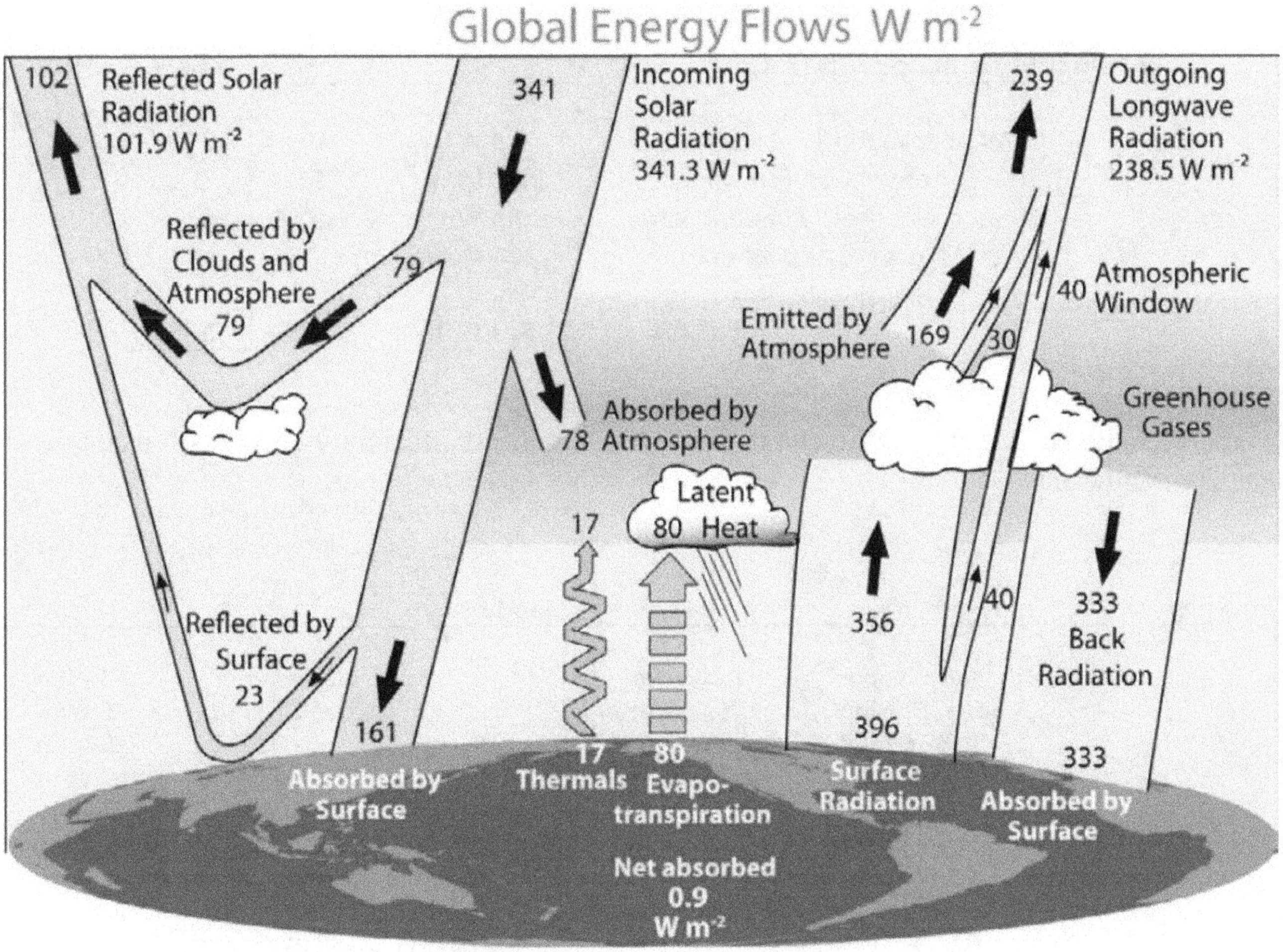

Illustration 120: Unmodified KT08 energy balance. It shows the Earth's surface losing 306% more energy to the atmosphere than it gets from the Sun. This misdirection is accomplished by using radiative flux instead of energy transfer.

62% Cloudiness:

This section heavily depends on the KT97 paper. That is the paper that spent the most time on the issue of clouds and the energy balance. The topic is a challenging one and it took several iterations until I had what I felt was the most accurate result possible. The 62% cloudiness is the value used in KT97 as the normal and that is why I chose that value.

So different is the behavior between cloudy and clear conditions (to say nothing of the in-between) that there are essentially two separate behaviors that are then averaged based on the average amount of "cloudiness" for the Earth. This methodology is reasonable, but the result is an estimate and far from a precise measurement. Clouds also affect the rate of convection as the behavior on the surface is different when it is cloudy compared to clear skies.

That the average amount of energy absorbed by greenhouse gases is 23 W/m² is enough information to show how different the atmosphere behaves when comparing cloudy to clear sky conditions. The overall energy balance for the atmosphere is:

Atmosphere Absorbed (AA)			%	TSI %
SW Absorbed (Stratosphere)	78.0 W/m²		39%	23%
Convection (Thermals)	17.0 W/m²		9%	5%
Evaporation	80.0 W/m²		40%	23%
Greenhouse Gas Absorbed	23.0 W/m²		12%	7%
Total AA	**198.0 W/m²**		**100%**	**58%**

The greenhouse gas absorption can be broken down to the impact by each gas for cloudy and clear sky conditions.

Clear Sky			Cloudy Sky		
Water Vapor	13.8 W/m²		Water Vapor	20.7 W/m²	
Carbon Dioxide	6.0 W/m²		Carbon Dioxide	1.6 W/m²	
Ozone	1.8 W/m²		Ozone	0.5 W/m²	
CH4 + N2O	1.4 W/m²		CH4 + N2O	0.2 W/m²	
Total	**23.0 W/m²**		**Total**	**23.0 W/m²**	

When the different conditions are considered for the accepted 62% cloudiness of the Earth, the result is the estimate that I used for the energy absorbed by the Earth's greenhouse gases.

62% Cloudiness			%	TSI %
Water Vapor	18.1 W/m²		79%	5.3%
Carbon Dioxide	3.3 W/m²		14%	1.0%
Ozone	1.0 W/m²		4%	0.3%
CH4 + N2O	0.7 W/m²		3%	0.2%
Total 62% Cloudiness	**23.0 W/m²**		**100%**	**6.7%**

The range for CO_2 in the RHT to the atmosphere is 7-26% of the energy transferred to the atmosphere. That lines up very accurately with the values usually thrown around for the role that CO_2 plays in the overall greenhouse effect. The error in that line of thought is that greenhouse gases account for only 19% of the total GHE. So the 7-26% is really 1.3-4.9% of the GHE. They

198

are in fact overstating the effect of CO_2 by a factor of 5.

That is independent of the cloudiness. The cloudiness just weights the values between the clear and cloudy conditions. If the cloudiness changes, then the impact of CO_2 will slide between the 1.3-4.9% of the GHE. A 10% increase in the impact of CO_2 will not be significant for reasons to be explained later in the book.

Chapter 12 The Earth's Atmosphere

It was while I was researching the Earth's energy balance that it became clear to me that there appeared be a real problem with the theory of global warming. CO_2 plays a part, but only a small part of the overall energy balance. Much like CO_2 plays a small part of the energy balance, it plays an even smaller role as an atmospheric gas. To understand the possibilities, a better understanding of the atmosphere is required.

The energy balance and the Earth's atmosphere are intimately entwined. A discussion of one is impossible to have without the other being part of it. The nature of the atmosphere influences the energy balance. The atmosphere is changed by the energy it gains. That is the nature of a feedback system. A change in one causes a ripple of changes that go back and forth.

Water vapor is the best example of a feedback. Increasing the amount of water vapor in the atmosphere causes many changes. It could increase the number of clouds which increases the energy that is reflected away from the Earth, but it also will reduce the amount of energy that escapes the Earth. If there were any greenhouse gases that could cause the Earth to warm it is water vapor[130]. That is why the theory of global warming depends on a water vapor feedback[131]. The proposed feedback is that an increase in the amount of CO_2 will cause an increase in the amount of water vapor in the atmosphere. A majority of the theorized temperature increase is actually supposed to be caused by water vapor.

Understanding if that is possible requires an understanding of the atmosphere. It is through feedbacks that the energy balance and the atmosphere become so entwined. A change to a part will cause other changes to take place. Understanding the atmosphere is an important part of the whole story.

The atmosphere is what truly makes the life on the Earth's surface possible. Without the atmosphere it might be possible for some of the life in the sea, but it would be a very different type of life. As it is, almost all life on Earth depends on the atmosphere in some way. Plants depend on the atmosphere to provide the carbon they need to grow. Without the CO_2 in the atmosphere, they would die as quickly as the animals would die without the O_2 that the atmosphere provides. The entire exchange of gases between plants and animals depends on the atmosphere to properly mix the gases so they are readily available to all life on Earth.

The atmosphere also acts as a storage unit for the gases that are important to life on Earth. The Earth's atmosphere is stable as a result of this. The atmosphere has many sources for the gases that are always being used. Oxygen is critical for life, but it is not a naturally stable gas in the atmosphere and would diminish to nothing in only a few thousand years if it was not always being replenished[132].

The storage capacity also applies to the plant kingdom. In the NH spring, the surge of plant growth that happens each spring removes ~6% of the CO_2 that exists locally in the atmosphere. The drop is so large on the large landmasses of the NH that the entire global average drops when the plants start growing each spring. The yearly chart for CO_2 in North America from 2006 is typical. In 2009 the Park Falls, Wisconsin data shows that the CO_2 level dropped from 395 to under 370 ppm from April to August. That is a 6.5% drop in the CO_2 level that is directly caused by the plants pulling the gas from the atmosphere.

The emperor penguins in Antarctica depend on oxygen that is formed thousands if not tens of thousands of miles away, but carried by the atmosphere where it can be readily used in the depths of the winter.

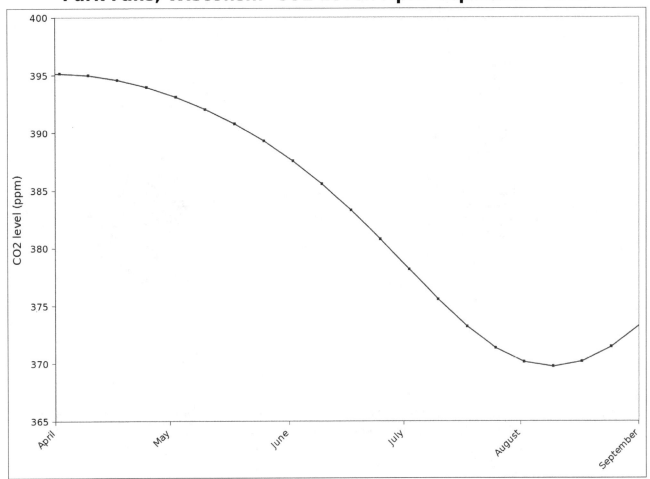

Illustration 121: The CO_2 level in the middle of North America follows this seasonal pattern.

Having a basic understanding of what the Earth's atmosphere is composed of and how it behaves is needed. Only then can the seasonal changes to the insolation and atmosphere be combined to explain why the Earth's temperature behaves the way it does over the course of a single year. Without an understanding of a single year it is simply not possible to predict longer periods of time.

The first step is to learn what the atmosphere is made of and how each gas impacts the Earth's temperature. Each of the main 5 gases play an important role in the atmosphere much like each flow of energy is an important part of the overall energy balance. Hopefully by explaining each gas and how it interacts with the energy balance as a whole it will be clear why the Earth's climate behaves the way that it does.

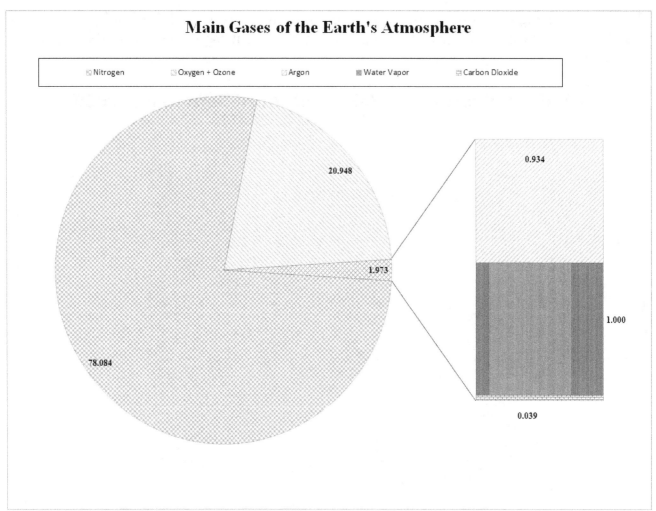

Illustration 122: There are more gases than these, but these above gases are the most critical for the Earth's climate and they also compose greater than 99.99% of the atmosphere. Technically the percentages here add to over 100% because water vapor is included.

This is an overview of all the gases and it also includes water-vapor which is normally not considered a gas in the atmosphere, even though it is ~1% of the atmosphere by volume. The percentages of the other gases are for a dry atmosphere without water-vapor being included. I have also included ozone with oxygen because there is so little ozone in the atmosphere that it would simply not show on the chart the overall concentration is so low, but it does play an important role.

I will discuss each gas here in more detail. Each of these gases plays a role that needs to be understood. I will also provide an estimate of how much of a role each gas plays in the Earth's GHE. Without the atmosphere the Earth would be much colder, the real question is how critical is each gas to keeping the Earth warm?

Nitrogen and Argon:

Nitrogen is by far the most common gas in the atmosphere. For every 1,000,000 atoms of gas in the atmosphere it makes up 780,840. That means that it makes up 78% of the Earth's atmosphere.

Argon is the 3rd most common gas in the atmosphere and it behaves in a similar way to nitrogen which is why I am putting it with nitrogen. This is a normal approach to dealing with the inert gases in the atmosphere. Argon makes up 9,340 atoms of every million. Together they make up 790,180 ppm (79.0%) of the Earth's atmosphere[133].

These gases are for all purposes the inert portion of the atmosphere. They do not directly absorb energy from the Sun or from the surface of the Earth, but they are still important to the overall Greenhouse Effect. This is because gases in the atmosphere tend to be well mixed which means that all gases have the same temperature and concentration from one section to the sections in the same vicinity of each other. It isn't possible to warm up only a single gas in the atmosphere, they all have to warm. Nor is it possible to separate one gas from the others, at least not without a large amount of work.

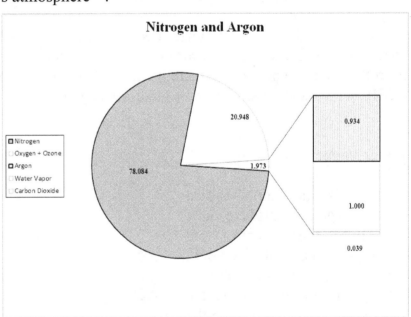

Illustration 123: (Purple) Nitrogen, (Yellow) Argon make up ~79% of the atmosphere.

So for the atmosphere to warm up, the inert 79% of the atmosphere must be dragged along. The 79% of the atmosphere that is inert plays an important role in *reducing* the fuel efficiency of everything from cars to coal power plants. This is because when something is burned (gas, coal, wood or anything else), much of the energy that is released during combustion must go into warming up the inert gas that is mixed with the oxygen.

This is why welders use tanks of pure oxygen to produce a hotter flame. Using air limits the temperature of the flame because of the energy that is used to warm up the nitrogen and argon that exist in the air. Having tanks of pure oxygen for your car would increase the fuel efficiency of any car, but of course pure oxygen is expensive and it is far easier to simply accept the reduced fuel efficiency that results from using air.

The second aspect is that the inert gases are still warmed by the surface of the Earth. Much of the convection energy is absorbed directly by the inert gases. This happens simply because the atmosphere is always in contact with the surface of the Earth. An atmosphere with only nitrogen and argon would still absorb just as much energy by convection as it does with the other gases in the atmosphere. So the GHE would still exist to some degree even if there were no oxygen, water vapor or CO_2 in the atmosphere.

Since 14% of the energy balance from the surface to the atmosphere is convection, almost 5 °C of the 33 °C would take place without the presence of any greenhouses gases at all. An Earth with only nitrogen and argon would be cooler than it is today, but the Earth would still experience an atmospheric GHE without any greenhouse gases in the atmosphere. All that is required is sufficient gas in the atmosphere to absorb energy from the surface.

The method of energy transfer does not matter. In a nitrogen/argon only atmosphere there would be no infrared energy absorption. The atmosphere would still gain large amounts of energy from the surface, enough that there would still be a significant GHE. What really gives the Earth a GHE is the most important of the greenhouse gases.

Water Vapor:

If the Earth had a nitrogen/argon atmosphere only, but had the oceans and geography it had today it would not take long for water vapor to fill up the lower part of the atmosphere. That is because the amount of water vapor in the atmosphere is determined by the temperature (ignoring pressure since it is constant enough for the lower troposphere).

The temperature of the air determines how much water vapor the air can hold. The temperature of the water

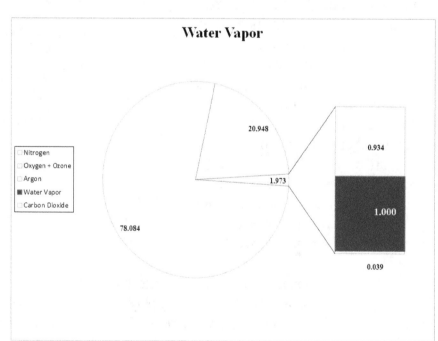

Illustration 124: (Dark Blue) Water vapor is not normally included as part of the atmosphere, water vapor dominates the behavior of the atmosphere at ~1%.

indicates how much water vapor will evaporate. If the water is warm and the air is not, then water will evaporate and then condense in the air above the water. This is what causes morning mist on cool mornings. This transfers energy from the warm water to the air above the lake. When the air is warm in the afternoon, this effect does not happen. It is very specific to cool air over warmer water.

Illustration 125: Morning mist is an example of energy transfer from the surface to the atmosphere. It is the most important component of the Greenhouse Effect.

That is the most visually recognizable form of the atmosphere absorbing energy from the surface. Even when it cannot be seen because the air is too warm to cause the water vapor to condense right away, it is the most dominant source of energy transfer to the atmosphere.

Water vapor is not usually considered one of the gases of the atmosphere even though it makes up ~1% of the atmosphere[134]. That is because it does not act like it is part of the atmosphere. It is a short-term visitor and most estimates are that a water molecule is only in the atmosphere for ~10 days[135] before it falls back to the surface in the form of precipitation.

Not only is it a short-term visitor, it varies a great deal by region, something that the respectable atmospheric gases do not do. Over the warm tropical oceans water vapor makes up 4% of the atmosphere. In those regions it is the 3rd most common gas in the atmosphere behind nitrogen and oxygen. Much of the rainfall around the world depends on the high concentration of water vapor that exists in the tropics. Large currents of moist air with low density (higher water vapor and temperature both reduce density) flow away from the equator to the higher latitudes.

The farther away from the warm oceans the less water there is in the air. The extreme would be in Antarctica during the winter. There the concentration of water in the air is ~0.004%

(40ppm). That makes it the 5th most common gas in the atmosphere. In the Antarctic winter water vapor plays a very small role in regulating the atmosphere.

In addition to regional differences in the concentration of water vapor, it also varies with altitude. Once above 10km (33,000 ft.) there is very little water vapor in the atmosphere. There are several reasons for this, but it comes down to the fact that water vapor turns into ice or small drops of water as the temperature decreases. This results in the water vapor turning into precipitation which is how it gets back to the Earth's surface.

No other gas in the atmosphere behaves in this way. It is for this reason that it is generally not included as one of the gases in the atmosphere. Most percentages of the atmosphere include water vapor as an asterisk at the end of the table. Water vapor is simply too abnormal in its behavior to be treated like the other gases.

The same things that make water vapor behave so differently are the same things that make it the most important gas for regulating the Earth's temperature. It is single-handedly responsible for ~82% of the total GHE. If the full GHE is 33 °C, then water vapor is responsible for 27 °C of that. There are two parts to this. First is the latent heat (evaporation) transfer to the atmosphere (66.7% of the GHE) and the greenhouse gas absorption by water vapor (15% of the GHE)[136].

Not only that, it is responsible for much of the energy that is reflected away from the Earth. The clouds as well as the ice, snow and surface water all reflect energy away. If water did not reflect away the energy that it does, the surface would absorb much more energy and the Earth would be warmer.

So if the Earth had only nitrogen, argon and water vapor, the temperature of the Earth would be almost identical to what it is today. Water vapor plays such a dominant role in regulating the Earth's climate that it is almost solely responsible for all the energy that is transferred from the surface of the Earth to the atmosphere. If the natural convection that would take place with any gas is included, the total GHE of the Earth would be almost 32 °C. This means that with only the energy transferred by: convection to nitrogen/argon, the latent heat of evaporation and water vapor absorption of infra-red, the average temperature of the Earth would only be ~1 °C cooler than it is today.

Oxygen and Ozone:

Oxygen is plentiful in the atmosphere at 209,480 ppm. This makes oxygen the 2nd most common gas in the atmosphere. Oxygen and nitrogen together account for 99% (990,320 of every 1,000,000 atoms) of the atmosphere. If the noble gas argon is included then 99.966% of the atmosphere is accounted for. This is why all other gases are considered trace gases.

Ozone is so rare in the atmosphere as a whole that it doesn't even register. Even in the ozone layer (30 +/- 10km) the concentration of ozone is very low at 8 ppm[137]. This is due in large part to the highly reactive nature of ozone. If the ozone was not constantly being regenerated in the atmosphere it would cease to exist in less than 9 days[138]. This has nothing to do with pollution and is simply due to the highly reactive behavior of ozone.

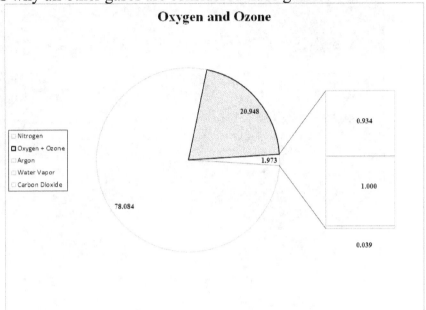

Illustration 126: (Light Blue) Oxygen is ~21% of the atmosphere. Ozone is included, but does not even have a concentration of 1 ppmv.

Ozone is formed when oxygen is hit by ultraviolet light from the Sun. The interaction of the UV and oxygen create 400 million tons of ozone every 24 hours. That is ~12% of all the ozone in the atmosphere. The energy involved in creating 400 million tons of ozone is significant. It is that energy that causes the stratosphere to be warmer than the upper troposphere.

Ozone is also the most reactive gas in the atmosphere. In normal atmospheric conditions the concentration of ozone will drop by 50% in only 30 minutes[139] as it reacts to form regular oxygen. This high reactivity is why it is a pollutant in the lower atmosphere. In the stratosphere where the ozone layer resides, the rate of formation by UV energy is in balance with the rate of decomposition.

It is the oxygen/ozone cycle that causes most of the energy absorption by the atmosphere from the Sun. The warming that takes place in the atmosphere as a byproduct of the cycle is limited to the upper atmosphere. By this measure oxygen and ozone absorb far more energy than any other atmospheric gas. The difference is that they absorb the shortwave (SW) energy from the Sun instead of the longwave (LW) energy from the Earth's surface. Since this energy has little impact on the temperature of the Earth's surface[140] it isn't necessary to include it with the energy balance of the lower atmosphere that interacts with the Earth's surface.

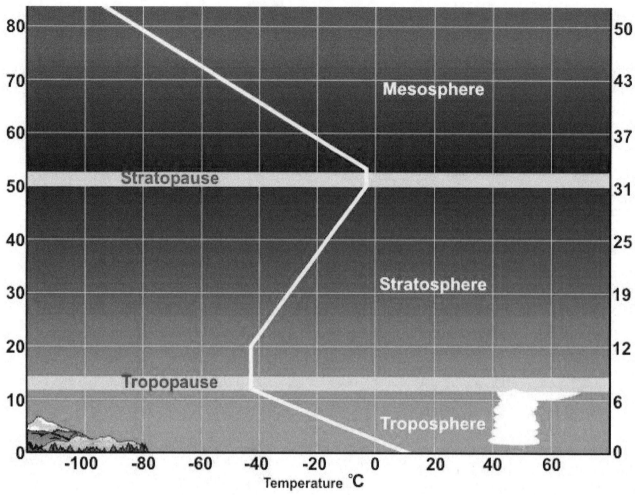

Illustration 127: The Troposphere is the part of the atmosphere in which we live. Most of the atmosphere's mass is within the Troposphere. The Stratosphere is warmed by the ozone/oxygen absorption of UV from the Sun.

Ozone and oxygen do absorb a small amount of LW energy from the surface. The total amount is very small at only 1 W/m² which accounts for 0.8% of the total energy transferred from the surface to the atmosphere. That percentage of the total GHE would be ~0.3 °C, but since ozone is limited to the upper atmosphere, that warming has little impact on the GHE.

This leaves only one more significant gas in the atmosphere. That final gas is the one most of you have been waiting for. It is of course, the notorious carbon dioxide.

Carbon Dioxide (CO_2):

CO_2 is the 2nd most important trace gas in the atmosphere after ozone. CO_2 is the 4th most common gas in the atmosphere (if water vapor is excluded) and there is ~1,000 times more CO_2 than there is ozone in the atmosphere, but if ozone is given the full credit for blocking the UV energy, then ozone absorbs 24 times more energy than CO_2 does.

208

The amount of CO_2 has been steadily increasing for as long as scientists have been measuring the amount of CO_2 in the atmosphere (since the 1950's). The increase in CO_2 is believed to be caused by the emissions of combustion when a carbon source is burned with oxygen in the atmosphere to provide energy. While the concentration of CO_2 has significantly increased as a percentage of the amount of CO_2 that existed 100 years ago, the amount of CO_2 as a percentage of the atmosphere has changed so little that in most cases it would be considered a rounding error.

The current concentration of CO_2 in the atmosphere is ~390 ppm. The amount that is considered natural is ~300 ppm. So the percentage of the atmosphere that is CO_2 has changed:

$$300/1,000,000 = 0.03\%$$

$$390/1,000,000 = 0.039\%$$

The increase in CO_2 over the past 100+ years has managed to change the amount of CO_2 from 0.030% to 0.039%. Even detecting that change requires very sensitive equipment. To put that 0.039% into perspective, it is like comparing the population of Topeka, Kansas to the entire population of the United States. That is 124,000 people in comparison to the total population of ~307,000,000.

Despite this, it is still the 2nd most important gas that absorbs infra-red energy. While it makes up 0.039% of the atmosphere, it is responsible for ~3% of the energy that the atmosphere absorbs from the Earth's surface. So while CO_2 is a small part of the atmosphere, it is a strong greenhouse gas. This leads to the question of what will happen as the concentration of CO_2 continues to increase. This needs to be understood as there is every indication that the level of CO_2 in the atmosphere will continue to increase.

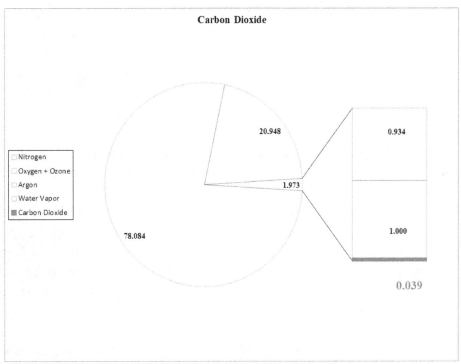

Illustration 128: (Red) CO_2 accounts for ~2% of 2% of the Earth's atmosphere.

Explaining the Greenhouse Effect

Dr. William Harper is a physicist at Princeton. He came up with what I consider the best analogy to explain the effect of the Earth's atmosphere and keeping it warm. His idea was specific to CO_2, but I will expand on it here. The basic idea is that the GHE acts the same way on the Earth's surface as winter clothing does on a person during the winter.

In cold weather, people dress appropriately to keep themselves warm. If I had to go outside to shovel the driveway when the temperature was -20 °C (-4 °F), I would dress for cold weather. If I didn't, I would lose too much energy to the cold air outside. I would feel cold because my body would be losing heat very quickly to the cold air around me. Our bodies sense the loss of heat and let us know when we need to put something on.

The first thing I would put on when the weather gets cold is a coat. That is what keeps a person from losing a dangerous amount of warmth in cold weather. Water vapor is the coat that is the key to staying warm. Without a coat, the largest and warmest part of my body is losing heat to the cold air. That is why coats are the most commonly worn item when the weather turns cold. Coats are critical in preventing your body from losing too much heat.

Since I will be using my hands outside, I would also put on gloves. When hands are in contact with the cold air they also need to be covered. Gloves are not critical to keeping your body warm, but they are critical to keeping your hands warm. I would not get hypothermia from not wearing gloves, but my hands could get very uncomfortable. Gloves are equivalent to the convection or thermals that would happen with any substantial atmosphere. They are not critical, but they sure make things nicer.

If it were windy there is one more thing I would put on to stay warm. The less hair a person has, the more important it is to wear a hat. This is what Dr. Harper compared CO_2 to when he testified before the US Congress on February 25, 2009[141]. The Earth is warmer because it has CO_2, but it is not critical to keeping the Earth warm. An Earth with 300 ppm of CO_2 is an Earth that is wearing a warm hat. Doubling the amount of CO_2 in the atmosphere from 300 ppm to 600 ppm would be exactly like changing from a warm ski hat to a slightly better ski hat. Your head will be slightly better insulated, but it will certainly not make that much of a difference. The difference between having a hat and not having a hat matters far more than having a good ski hat and having a better ski hat.

A person going outside in -20 °C would lose ~400W of energy to the cold air. The purpose of insulating winter clothing is to reduce the amount of heat loss to a level that a person is comfortable with. Most people would consider 25 °C (77 °F) to be a comfortable temperature. The radiative heat loss of the human body at that temperature is ~100W. Good winter clothing does not prevent heat loss (that is basically impossible), it reduces it so your body is losing heat at the same rate it would if it were in a warmer environment. So think of winter clothing as a way of simulating a warmer environment for your body.

Because the human torso has the most surface area and is the warmest part of the body, it loses the most heat. That is why a coat is the most critical part of keeping your body warm. Of the 300W reduction in heat loss, a coat is responsible for 270W (2.7 100W light-bulbs) or 90% of the reduction. Adding a hat prevents a total loss of ~12W. Switching to a better hat could reduce the heat loss from 300W to 301W. That is a change of 0.5%. A value so small as to be more theoretical than having any meaning.

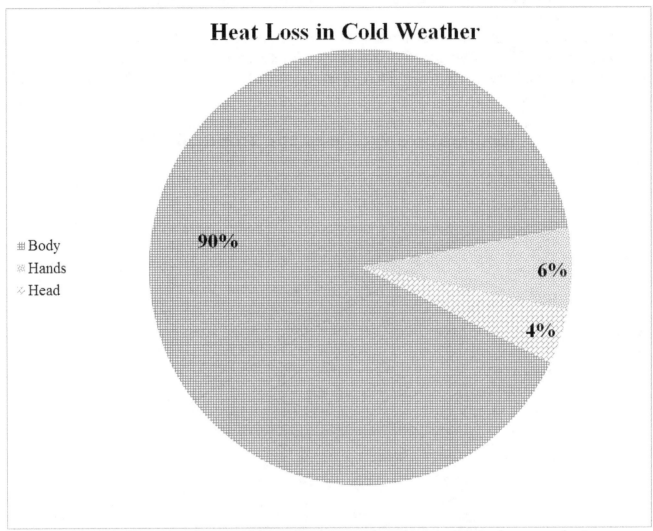

Illustration 129: Wearing a hat in cold weather is a good analogy to CO_2 in the atmospheric energy balance. Changing to a better hat is what increasing CO_2 does.

Since the purpose of winter clothing insulation is to make you feel like you are in a warmer environment it is possible to precisely determine the temperature difference it would make to your body. Instead of feeling like you were in a room that was 25 °C, it would feel like you were in a room that was 25.2 °C.

The reason that these objects make a difference is that they insulate your body. Insulation works because it conducts temperature very slowly. So the inside of a coat is the same temperature as your body and the outside is the same temperature as the outside air. In-between those two surfaces the temperature will be linear between the temperature of the two surfaces. The better an insulator is, the less heat it will transfer across a large temperature.

A good coat will be thin and not lose heat quickly even though there will be a 50 °C temperature difference between the inside and outside of the coat. The atmosphere behaves exactly like an insulator between the surface of the Earth and the top of the troposphere. The temperature profile of the troposphere is strongly related to the altitude above the Earth's surface.

At the surface the temperature is an average of 14 °C. At the top of the troposphere the average temperature is -40 °C. That 55 °C temperature difference is comparable to the temperature difference that a coat deals with when the temperature is -20 °C. The atmosphere is the insulator for the Earth's surface.

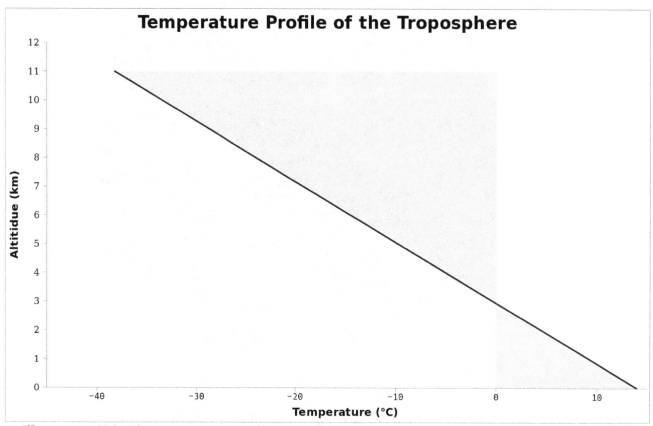

Illustration 130: The temperature profile of the Troposphere is the only proof required to show that energy is flowing from the Earth's surface to the top of the Troposphere. Fourier's Law is that a constant flow of heat (from hot to cold) is required to maintain a linear temperature gradient.

It allows the Earth to lose energy only at the rate that energy is transferred to the atmosphere. Much like a coat keeps your body warmth in by limiting how quickly energy transfers through the coat. If the Earth's atmosphere disappeared then it would lose energy to space at a much higher rate until the surface cooled down enough to be in a new and much colder equilibrium with space.

Adding more CO_2 to the atmosphere will slightly increase how effective an insulator the

212

Earth's atmosphere is, but since the amount of energy that CO_2 is responsible for is so small already, the effect of doubling the amount of CO_2 will be even smaller. It would be very similar to the effect of switching to a slightly better ski hat.

The current energy balance shows that CO_2 is responsible for ~3.3 W/m^2 of energy to the atmosphere. Doubling the concentration will increase that to ~3.7 W/m^2. The impact that would have on the total GHE would be ~0.1 °C increase in temperature. Since the CO_2 level has increased only 30% of a doubling in the past 100 years, the current impact of increased CO_2 levels in that period of time is ~0.03 °C.

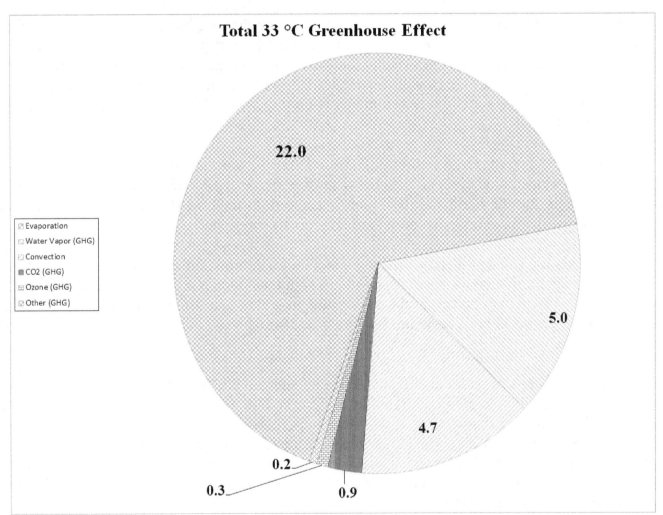

Illustration 131: The total Greenhouse Effect is a function of how much energy is transferred to the atmosphere. This is a summary of the parts that transfer energy to the Earth's atmosphere.

This explains why elevated CO_2 levels at the end of the Eemian failed to prevent a glacial period from developing. While CO_2 is the 2nd most important greenhouse gas, overall it plays a very small role in the total GHE. This is not to say that the GHE is not important, because it is critical to life on Earth. Without the atmosphere the Earth would have a temperature that is known as the blackbody temperature.

Blackbody Temperature: *The theoretical temperature of the Earth if it had no atmosphere. The value is -19 °C (254K)[142].*

The blackbody temperature is explained in detail in the science section and I will leave it there where it belongs. The overview is that because the current average temperature of the Earth is 14 °C (287K) and the blackbody temperature of the Earth is -19 °C, the estimated greenhouse effect is 33 °C[143]. Warmists generally argue that almost all of the 33 °C is caused by greenhouse gases.

Without an atmosphere, the Earth would have a climate that is similar to the Moon's. In fact the average temperature of the Moon is comparable to the blackbody temperature of the Earth[144]. How the temperature behaves on the Moon around that average temperature is very different from how the Earth behaves. The atmosphere moderates the Earth's climate. It does so simply because it gains energy from the surface and that energy is used to warm the atmosphere and the ground beneath the surface.

As a result the daytime temperatures on Earth are much lower than they are on the Moon, but the night time temperatures are much higher than they are on the Moon. The amount of energy per square meter that the Earth and the Moon receive from the Sun is identical, but in the end they have very different climates because the Earth has an atmosphere.

In the Scientific Content there is also some additional information as to why the 33 °C is not all caused by the Earth's atmosphere. Much of it is, but that value is a bit of an overstatement. It could be more accurately stated as the maximum possible value of the GHE is 33 °C. For practical purposes the 33 °C is accurate enough and that is the value that I will use as the baseline greenhouse effect.

There is one problem with the GHE that must be addressed here. I have already shown how the Earth's temperature varies over the course of the year, but this also proves that the GHE is not a constant based on the atmosphere. If the greenhouse gases caused the GHE, then it would be constant over the course of the year. That is not the case. As always, every facet of the Earth's temperature, including the GHE itself, is determined by the season that the Northern Hemisphere is experiencing. Yet warmists continue to ***deny*** that the regional insolation matters. It makes me wonder who is really anti-science.

214

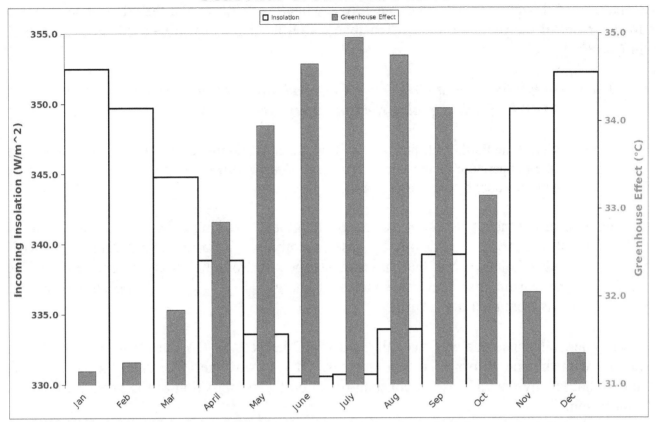

Seasonal Greenhouse Effect

Illustration 132: Saying the Earth has a 33 °C greenhouse effect when the actual total varies proves that the effect is not a constant based solely on the Earth's atmosphere.

The annual 4 °C swing in the actual temperature of the Earth shows that the GHE is not a constant 33 °C. If the GHE was caused solely by greenhouse gases, then the GHE would be constant over the course of the year because the atmosphere of the Earth does not show a seasonal variation that would explain this behavior of the GHE.

The Earth's GHE is in sync with the seasons of the NH. That cannot be the case if the GHE is dependent on the greenhouse gases in the atmosphere. The only explanation for the seasonal behavior of the GHE is the importance of the 65N insolation. When 65N is in winter, the entire GHE of the Earth is reduced to almost 31 °C. When 65N is in summer, the GHE of the Earth reaches it's current seasonal maximum of 35 °C.

Even though the Earth is getting an additional 24 W/m^2 of energy from the Sun, the Earth has a GHE that is 4 °C lower. Not only that, but July is when the CO_2 level in the NH is also reaching the lowest point of the year. Northern Hemisphere CO_2 levels in July are ~6.5% lower than they are in April[145].

The Earth has the highest GHE while CO_2 levels are near the yearly low and the Earth is getting the least energy from the Sun.

The very behavior of the Earth itself points away from the idea that the level of CO_2 in the atmosphere is what regulates the temperature of the Earth. Physical *observations* of the Earth itself show that the Earth's GHE is not caused only by greenhouse gases.

To me, one of the most frustrating aspects of the global warming debate is this often quoted idea that the entire GHE is caused *only* by gases absorbing[146] infra-red energy from the Earth's surface. That idea is false, but that has not stopped it from being used repeatedly. The atmosphere cannot distinguish between sources of energy. It is warm only because of all the energy that it gains from the Earth's surface.

That energy balance changes over the course of the year as the Earth's geography that is interacting with the energy from the Sun changes. That is why the GHE changes with the NH seasons. That is why the GHE is not constant based on the composition of the greenhouse gases in the atmosphere.

Observations are what prove scientific theory. Models are useful only when they match real world observations. In this case those real world measurements do not match the model of the Earth's GHE. This leaves the question: what does cause the GHE to change with the season of the NH? The only explanation is that the NH and SH respond differently to the Sun's energy.

It is convenient that this can be proven. The implications of the proof are profound and help explain things about the Earth's climate that are not yet fully understood. The next chapter will focus heavily on the implications of the proof, but for now I will just bring forward the proof that the energy balance of the Earth changes with the seasons of the NH.

The Changing Energy Balance

For as long as satellites have been measuring the temperature of the Earth, scientists have been able to measure how much energy is escaping the Earth's atmosphere. This energy that is escaping the Earth is called Outgoing Longwave Radiation (OLR)[147]. This is the energy that is transmitted by the atmosphere to space. Once this energy is gone, it no longer has the ability to influence the Earth's temperature.

Would it be surprising to find that the amount of energy leaving the Earth depends on the season of the Northern Hemisphere? At this point in the book it should not be a surprise that almost all aspects of the Earth's climate depend on the season of the NH and the amount of energy that is escaping the Earth is no exception.

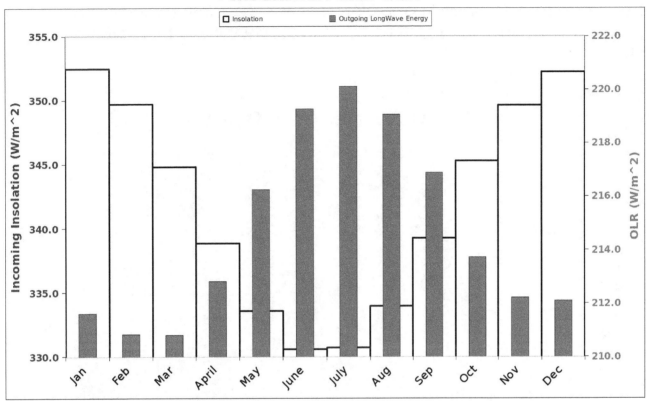

Illustration 133: The amount of energy leaving the Earth's atmosphere is not dependent on the amount of energy the Earth is receiving from the Sun. It also is dependent on the season of the Northern Hemisphere.

The cause of this behavior is critical and will be discussed in detail in the next chapter. What it proves is that the Earth's energy balance is dependent on the season of the NH. Each year the amount of energy leaving the Earth varies by ~10 W/m² and it is out-of-phase with the energy the Earth is receiving from the Sun. Why this happens will be discussed, but I will say now that the physical law that causes this to happen also proves that anthropogenic global warming is virtually impossible.

Before I get to that though, I need to discuss a few more things that are involved with the Earth's atmosphere and the global warming debate. As these are more involved with the global warming debate, they are not critical to understanding the Earth's climate, but they are important to the details of the global warming discussion.

The Two Ways to Make CO₂ Appear More Important than it really is

There are many ways in which data can be presented. Everyone shows data in a way that supports their arguments. I will freely admit that I have spent much of my time presenting the data in a way that causes the least confusion and supports my conclusions. When presenting complex information it is helpful to simplify what it presented. This happens a lot in the climate debate and I would like to point out some of the choices that warmists have made to present their own arguments.

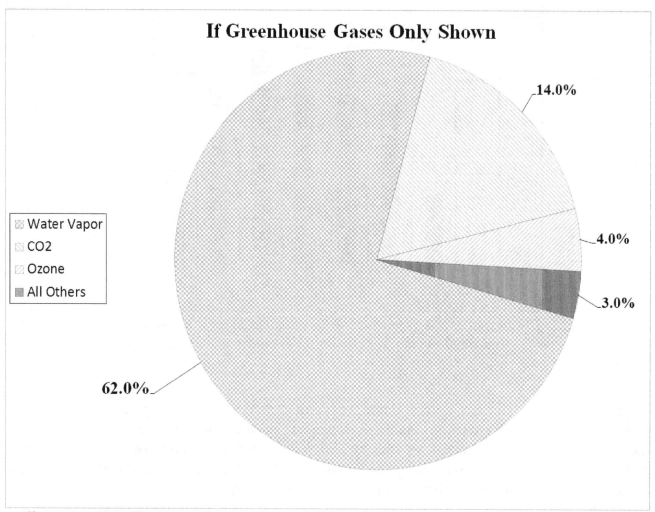

Illustration 134: CO₂ is significant to the total GHE only if the assumption is used that greenhouse gas absorption is the ONLY method of energy transfer. For 62% cloudiness this is the energy impact by greenhouse gas absorption only.

The most common one is to present only the greenhouse gases as the source of the GHE. In that way they can argue that CO₂ is responsible for 9-26% of the GHE (using the 62% cloudiness results in 14%).

When the piece of the pie is 14% instead of 1% it is easier to sell that a change will cause problems for the Earth's climate. This is by far the most common method of trying to show that CO_2 is more important than it really is.

The problem is that it isn't quite enough by itself to show that a change will cause the warming that is predicted. To make an even greater appearance of warming they switch to the Energy Balance discussed in the Scientific Content of the previous chapter. They don't use energy at all. They use radiative fluxes for greenhouse gases and compare that to the energy transfer by other methods.

Radiative Flux: *Infra-red energy that all physical objects radiate.*
See Science Content for more details.

Forcing: *The same thing, but used to describe the infra-red energy from clouds and gases. There is no practical difference, but infra-red from gases can be distinguished by the proper analysis. Radiative flux is the proper term for this type of energy.*

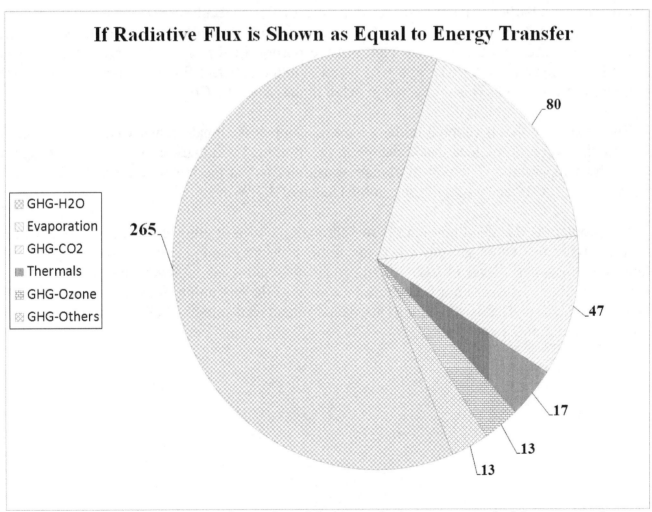

Illustration 135: This uses the ~radiative flux of the atmospheric gases instead of the amount of energy transferred as a result of each greenhouse gas. This is the best way to show that a change in CO_2 levels could have a measurable impact.

In the case where radiative fluxes are directly compared to energy transfers (like evaporation) the appearance of greenhouse gas domination appears. In this case alone, the doubling of CO_2 appears to cause a significant change in the energy balance, especially if it is assumed that an increase in the CO_2 will also cause an increase in the amount of water vapor in the atmosphere. A doubling of CO_2 will increase the amount of radiative flux (not energy transfer) by 3.7 W/m². That each doubling will cause an increase of 3.7 W/m² "forcing" is the foundation[148] of the theory of global warming.

In the Scientific Content there is a section on the impact of different radiant fluxes and what the effective impact is of those. The most important number for warmists and radiant "forcing" as they call it is 3.7 W/m². That is the increase in forcing if the concentration of CO_2 doubles. In the above case, the forcing of CO_2 would go up by 3.7 W/m² if the concentration doubled. This would change the "forcing" of CO_2 to a value of 50.7 W/m².

An increase in the radiative flux is real, but a change in a flux is not the same as a change in the flow of energy. Earlier I mentioned that doubling of the concentration of CO_2 would cause a ~10% increase in it's portion of the total GHE. It is the increase in the radiative flux that would cause that increase. So the effect is real, but the portion of the GHE that is impacted by the change in flux remains at 0.9 °C. So a 10% increase in the radiative flux would increase the GHE on the Earth by ~0.1 °C. This is covered in detail in the Scientific Content.

The final topic that is covered in the Scientific Content is climate sensitivity. This is a more detailed discussion of the idea that a change in the "forcing" will cause a proportional change to the Earth's temperature. The typical climate sensitivity is that the temperature of the Earth will increase by 3.0 °C if the "forcing" goes up by a value of 3.7 W/m².

The problem is that giant holes are blown through the idea of climate sensitivity at the end of every interglacial. Those periods when the Earth is experiencing dramatic long-term cooling while the atmospheric level of CO_2 remains at interglacial levels is always bad for the idea of climate sensitivity. It works much better for warmists at the beginning of interglacials than at the end of interglacials. That is why almost all glacial-interglacial studies of climate sensitivity focus on the beginning of the interglacials.

Here is a summary of the main topics covered so far in the book. These are basically in order from the beginning of the book and not necessarily in order of importance.

1.   The Earth is ALWAYS changing temperature.

2.   Those changes in temperature take place on many different time scales.
        a. Daily
        b. Yearly
        c. Short-term (100 year time scale)
        d. Long-term (> 1,000 year time scale)

3.   Shorter-term trends mask the longer term trends. Predicting yearly temperature trend based on a single day or week is foolishness.

4.   Long-Term trends are caused by Milankovitch Cycles. This is the glacial/interglacial cycle of the climate cycle.

5.   Interglacial periods end when the 65N solar insolation anomaly goes negative.
        a.   The Holocene (modern day) 65N insolation went negative 3,000 years ago.

6.   Since the 65N insolation anomaly went negative the range of glaciers in the NH have expanded their range halfway to the equator. Very few NH glaciers are more than 4,000 years old.

7.   What happens to energy from the Sun?
        a. 23% Reflected by clouds.
        b. 7% Reflected by surface.
        c. 47% Absorbed by surface.
        d. 23% Absorbed by atmosphere (mostly all in the stratosphere).

8.   What happens to the 47% of energy absorbed by the surface?
        a. 50% Transferred to atmosphere by Evaporation.
        b. 11% Transferred to atmosphere by Convection.
        c. 14% Transferred to atmosphere by Radiative Heat Transfer.
        d. 25% Transferred to SPACE by Radiative Heat Transfer.
        e. Any remaining converted by surface (photosynthesis and such).

9.   The gases that cause the 23 W/m^2 (14% by RHT) transferred by RHT.
        a. 79% Caused by water vapor.
        b. 14% Caused by carbon dioxide.
        c. 1%   Caused by ozone/oxygen.
        d. 1%   Caused by all other greenhouse gases.

10.   The total GHE is caused only by energy that is transferred to the atmosphere. This is

because temperature changes only when energy is transferred.

11.   Carbon dioxide accounts for only 2.8% of the total GHE.

It's time to go back to those settlers in Kansas. This is the equivalent information that I would present about the yearly cycle of the Earth's climate. I would really hope that they would listen. Placing too much importance on daily trends to predict what is going to happen months in the future is insanity. No one would be taken seriously who predicted the weather a month in the future based on what the temperature was yesterday.

So why are 100-year predictions taken seriously based on a 20-year period of warming that took place between 1980-2000? The only answer I have been able to find is that the warmists are the only ones that have presented a reason to explain why the temperature has been increasing. That is really the problem most people have, they cannot explain why the Earth was warming when it really should have been cooling.

The real answer is that there is always some amount of variability, but that doesn't provide a scientific cause and effect and people have gotten used to knowing the reason why something happens. Rarely are they content to just accept that it falls within the statistical noise that happens when we live on a planet that is always changing.

There are also critics of the Milankovitch Cycle. While it is best at predicting when interglacials end, it is not always as good at predicting when interglacials start. There have been times where energy went up and the Earth's temperature did not. 105,000 years ago the energy peaked higher than it did during the Holocene, but the Earth didn't enter an interglacial. Nor can it easily answer why the glacial/interglacial cycle changed from 40,000 year cycles to 100,000 year cycles one million years ago.

Surprisingly, the cause of the current short-term warming and the times where the Milankovitch cycle should have predicted an interglacial but one didn't happen, appear to be related. That is also related to why the Earth seems to have a central temperature that does change over very long periods of time, but on times scales of a few thousand years it can vary greatly from that central temperature.

The next chapter contains a genuinely new theory that would appear to answer all of those questions. Even if I am wrong about other aspects of the climate, which I have to accept as a real possibility (and is something that all scientists should accept), this new theory is one that could answer all of those questions. It also shows that global warming is basically impossible.

Warning:

Scientific Content!!!

First off, I would like to apologize for the length of that chapter. Splitting out the energy balance from the atmosphere was a difficult task. I hope that I did so in a way that made sense. I won't say in a way that will make people happy because that is not the goal. The goal is to explain the energy system of the Earth/atmosphere in a practical engineering method. That it greatly contradicts the Theory of Global Warming is just the effect that such an explanation has.

The bad news is that the Science Content of this chapter is about as long as the chapter itself.

Blackbody Temperature of the Earth:

Only two things need to be known to determine the blackbody temperature of the Earth. The amount of energy the Earth gets from the Sun and the albedo of the Earth. The first is the same as what was used in the energy balance of the Earth which is 341 W/m². The albedo the ratio of energy that is reflected away.

0 (no energy reflected) < 0.3 (Earth's albedo) < 1.0 (All energy reflected away)

So the Earth reflects away 30% of the energy from the Sun. This is consistent with the energy balance in the previous chapter where 102 W/m² are reflected away and 239 W/m² are absorbed.

From there the blackbody temperature of the Earth is determined by the Stefan-Boltzmann Law which shows the relationship between an object's temperature and the amount of energy that is transmitted by the object.

$$P = \sigma T^4$$
P = energy in W/m2
$$\sigma = 5.67\,e\,{-}8\ W/m^2K^2$$

Since the energy leaving the Earth must be equal to the energy that is absorbed, the outgoing energy is known to be 239 W/m2. Solving for the blackbody temperature of the Earth is then simple.

$$T = (P/\sigma)^{1/4}$$

The temperature is of course the 254K that is the blackbody temperature of the Earth[149]. This is a very basic solution to a much more complicated problem. For an ideal situation it might be true, but it is too simple for the Earth as a whole.

In order to show this, it is important to explain the assumptions of the ideal situation. The key assumption is that all energy that is not reflected away is immediately transferred away from the surface. This is called an equilibrium situation. If the Earth had one side facing the sun at all times and the Earth's surface was a perfect insulator and there was no atmosphere to warm up, then the equilibrium situation would apply, but none of those conditions are true.

If any of those conditions which are true for the Earth are in play, it pushes the Earth into a non-equilibrium situation. The effect is that the temperature will be different than what is predicted by the blackbody equilibrium conditions. The Moon[150] is "almost" a perfect example and it still varies from the predicted blackbody behavior.

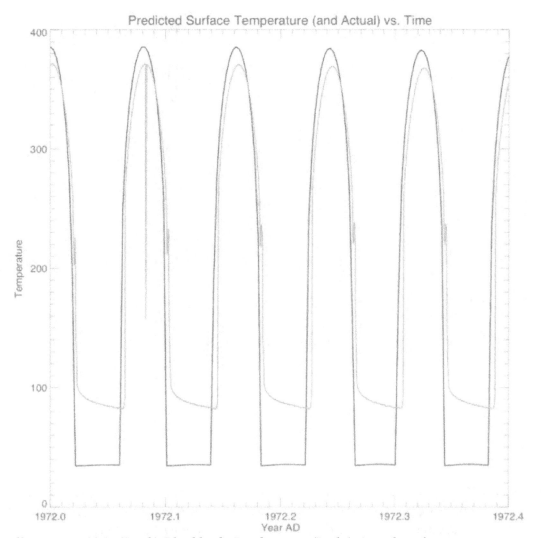

Illustration 136: (Dark) Blackbody Prediction. (Light) Actual surface temperature of the Moon in 1972.

NASA made the daily blackbody predictions for the Moon for the landings that took place more than 40 years ago and they found that the Moon's surface warmed up more slowly than predicted and also did not warm up as much as predicted. It also cooled down slowly and didn't cool down as much as predicted. This is exactly the effect that having an atmosphere has on the warming of the Earth's surface, but to a much lesser extent.

225

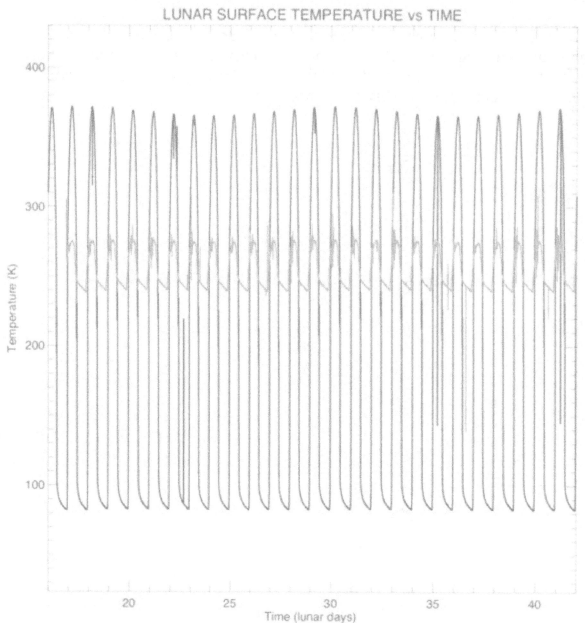

Illustration 137: (Light) Subsurface temperature. (Dark) Surface temperature from Apollo 15. The different high temperatures on the Moon are precisely in line with the distance from the Sun. January is closest to the Sun, so the daily temperature is higher.

The effect of cooler days and warmer nights than the blackbody calculations was a slightly warmer moon. The cause of this was the top 50 cm of the Moon's surface warmed up and stored energy from the Sun. The energy that seeped inward prevented the surface from warming up like a blackbody. After the Sun set, the reverse happened and energy from beneath the surface kept the surface from cooling.

In each of the three non-equilibrium conditions the Moon is closer to the equilibrium state than the Earth. The Earth conducts energy inward faster than the Moon, it rotates faster and the atmosphere also takes energy from the surface. The Earth also has oceans that cover 70% of the planet. They also change the situation a great deal.

The Earth is a true non-equilibrium situation and each day is proof of that. When the surface of the Earth is warmer than the air above it, net energy is transferred from the surface to the air above. When the air is warmer than the surface, the net transfer is to the surface. During the day the Sun warms up the surface (think hot concrete on a summer day) and that causes the air above it to warm. It also causes the surface underneath the Earth to warm up as heat is transferred underground.

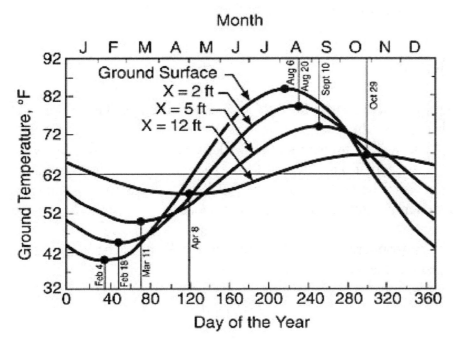

Illustration 138: Originally found at Virginia Tech, the website that hosted this is now gone.

During the night when the surface and the atmosphere are both cooling, energy flows from underground to the surface and then into the atmosphere. This effect alone is part of the GHE, but exactly how much is difficult to say. There are many types of geography and the ocean is also different. In general land surfaces in the NH show temperature change up to 3.5 m which is 7 times greater than the depth that the temperature changes on the Moon's surface. This indicates that far more energy goes underground on the Earth which also causes a portion of the observed GHE.

I won't venture a value for this effect in the total GHE, but it is clear that it does contribute to the observed GHE to some degree.

The GHE and the Eemian Interglacial

The climate cycle also shows that the GHE does not depend solely on the composition of the atmosphere. During the Eemian[151] there was a wide variety of temperatures, while the atmosphere was the same, including the CO_2 level. If the GHE was caused by greenhouse gases, then fewer changes in temperature would have happened. The GHE during the Eemian could have been as high as 40 °C, while having a lower CO_2 level than the Earth has now.

Since it is that correlation between the Earth's temperature and the level of CO_2 in the atmosphere that has convinced some people that it is the CO_2 level that is a main cause in the warming up and cooling down during the glacial/interglacial cycle. It would be bad for the theory of global warming if the GHE was shown to be higher with less CO_2.

Radiative Temperature of the Earth

It is also important to model how the Earth loses energy to space. The simplest solution for the Earth would have two parts, the energy from the surface to space and the energy from the atmosphere to space. The surface transmits energy like it is a blackbody at 287K, but because the atmosphere is only slightly cooler than the surface the net energy transfer by this method is small. The transmitted energy from the surface to space (40 W/m²) needs to be subtracted to determine the effective blackbody temperature of the Earth's atmosphere.

The energy that the atmosphere transmits to space is 198 W/m². The blackbody temperature of an object that that transmits energy at that rate is:

$$T = (198/\sigma)^{1/4}$$

$$T = 243K \ (-30 \ °C)$$

That is the temperature of the atmosphere at ~9.5km (31,350ft). It should be no surprise that is also the altitude where the amount of water vapor in the atmosphere starts to really drop off.

The upper reaches of the troposphere are where most of the Earth's energy is lost to space. The situation is far more complicated than this as well. Very large, hot and dry places (i.e. Sahara Desert) lose far more energy than places with high concentrations of water vapor. That is simply because without the water vapor in the atmosphere the energy tends to depart the Earth without getting absorbed by the atmosphere.

Cold places have lower temperatures so they lose energy at a lower rate. Places like Antarctica lose energy to space at roughly half the rate than hot, dry places do. The actual map[152] of heat loss from the Earth is far more complicated than the simple single temperature, but since using a single temperature works there is no reason not to use that simplification. The actual temperature of the

228

Earth and the effect that has on radiative heat transfer will play the key role in the next chapter.

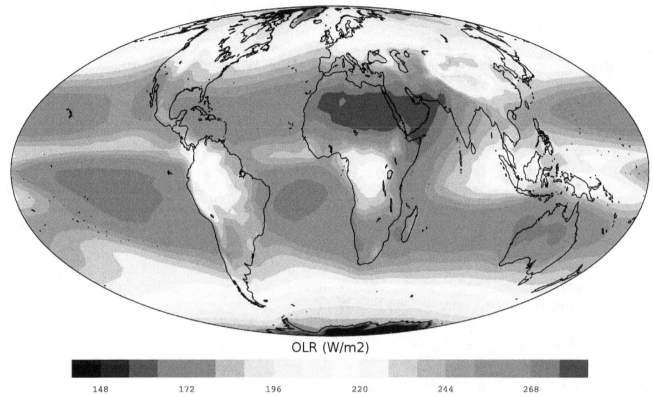

OLR (W/m2)

148   172   196   220   244   268

Illustration 139: OLR is highest in places with high temperature and low humidity. It is lowest in places with low temperature.

Temperature and Changing Radiative Fluxes

All radiative fluxes are dependent on temperature. The higher an object's temperature the more energy is transmitted. I am going to start with the example of a person radiating energy to the air around them. Instead of a frigid condition of -20 °C I am going to make it 20 °C. The person is still at their normal temperature of 35 °C (surface temperature, not internal). The radiative flux of the person and the air are:

Human Flux   =   511 W/m^2
Air Flux   =   419 W/m^2
Rate of Energy Transfer =   511 W/m^2 – 419 W/m^2 = 92 W/m^2

This means that a human being is transferring 92 W/m^2 for each square meter of surface area they have by means of radiative heat transfer (RHT). I am going to ignore the other transfers as they are not being debated.

Nowhere in that transfer of energy to the air does the concentration of the gases enter into the situation, but if I assume for a moment that it did, then the effect of doubling the concentration of CO_2 caused the flux to increase by 3.7 W/m^2, then the effect would be:

Human Flux	=	511 W/m^2
Air Flux	=	422.7 W/m^2
Rate of Energy Transfer	=	511 W/m^2 – 422.7 W/m^2 = 88.3 W/m^2

That would reduce the amount of heat that is transferred, but the effect would be identical to increasing the temperature of the air to 20.7 °C. So doubling the concentration of CO_2 would have an effect comparable to raising the temperature of the air by 0.7 °C.

The effect of RHT is temperature dependent, so using a human temperature and that of the surrounding air could be different than that of the Earth's surface and the atmosphere. The same type of analysis could be done using the radiative values from the KT08 paper from the previous chapter.

Surface Flux	=	396 W/m^2
Atmosphere Flux	=	333 W/m^2
Rate of Energy Transfer	=	396 W/m^2 – 333 W/m^2 = 63 W/m^2
Effective temperatures:		Surface: 16 °C, Atmosphere: 3.7 °C.

If the flux of 3.7 W/m^2 is added, then the result is:

Surface Flux	=	396 W/m^2
Atmosphere Flux	=	336.7 W/m^2
Rate of Energy Transfer	=	396 W/m^2 – 336.7 W/m^2 = 59.3 W/m^2
Effective temperatures:		Surface: 16 °C, Atmosphere: 4.4 °C.

Increasing the flux by 3.7 W/m^2 would result in an effective warming of 0.7 °C. So in this case the result did come out the same, but it cannot be assumed to be the case for all situations.

Most importantly, the surface is still transferring energy to the atmosphere. Regardless of anything else, net energy will always flow from the warmer object to the cooler object. In this situation the Earth's surface remains much warmer than the atmosphere and energy will continue to flow from the surface to the atmosphere.

This result is applicable to the Earth having a CO_2 level of 380 ppm, so doubling to 760 ppm would theoretically increase the Earth's temperature by an entire 0.7 °C. This is why the theory of global warming actually depends on water vapor feedback to cause most of the warming. That is the increase atmospheric CO_2 will also cause the amount of water vapor to increase. Most of the warming will actually come from elevated levels of water vapor in the atmosphere.

If I wanted to test this theory against history, I would go back to the Eemian interglacial and see how the Earth behaved. 125,000 years ago the Earth was warmer and the oceans levels were higher. All indications are that the water vapor level in the atmosphere would have been higher. The level of CO_2 was at 285 ppm (flux difference between 390 ppm and 285 is 2 W/m2, or 6%). All indicators are that the Earth in that state should have stayed warm.

The theory of global warming demands that the Earth in such a state would stay warm and possibly keep warming up. That is not what happened when the solar insolation decreased in the Northern Hemisphere. There was no significant change in the Total Solar Insolation (TSI), just in the specific location. When the Earth was in a state that it should have stayed warm according the theory of global warming, it instead entered glacial period that lasted for 100,000 years.

That the Eemian ended points to a significant flaw in the theory of global warming. It should not have ended. The Earth should have stayed warm or kept warming up. The only factor that could have caused the Earth to cool at the end of the Eemian was the change in NH insolation.

The Eemian is a true test of the theory of global warming. In the climate cycle each season must be compared to the same season. Comparing the temperature of the Earth in January to July would be foolishness because the position and condition of the Earth in each of those months are fundamentally different. The Eemian is the exact condition required to test the theory of global warming. Higher sea levels, less ice in Greenland and Antarctica. Warmer oceans and most especially a much warmer NH.

With all those conditions in place, what happened is the opposite of what the theory of global warming would predict. While the CO_2 level remained in the 280 ppm for 15,000 years from that very warm period of the Eemian 125,000 years ago, the temperature of the Earth dropped 10 °C over that 15,000 year period. The Earth went from 3 °C warmer than it is now to -7 °C. All while the atmospheric conditions for the theory of global warming held sway over the Earth.

None of that is a surprise if the analysis of the energy balance is on net energy transfer. Based on the energy transfer from the surface to the atmosphere, the role that CO_2 plays is small. Small changes in parts that play a small role are called insignificant. The role that CO_2 plays in the GHE is real. What is insignificant is the impact of CO_2 levels changing from 180 ppm during glacials to 390 ppm today. That 4.1 W/m^2 change in radiative flux is 12% of the total CO_2 flux. That total accounts for 1.0 °C of the GHE so the difference that CO_2 has made from the depths of the last glacial period to today is ~0.1 °C. The Earth varies 40 times that amount every six months. That might be a good way to define insignificant.

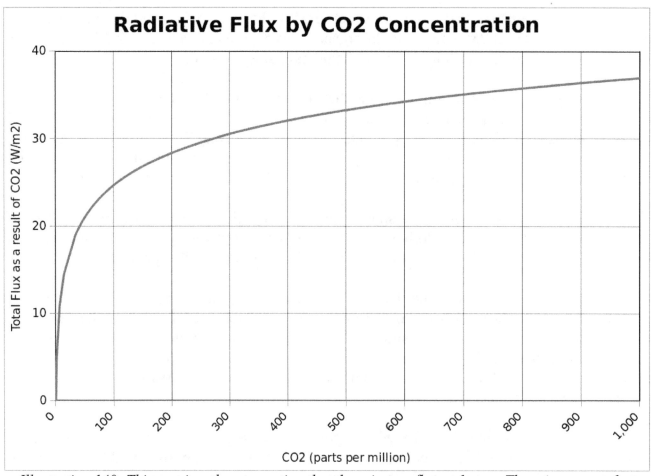

Radiative Flux by CO2 Concentration

Illustration 140: This requires the assumption that there is zero flux at 1ppm. There is some at that level, but for the sake of simplicity I call 1ppm the zero point.

Part of the reason that it is not a big deal is because of how the radiative flux of CO_2 changes with concentration. The effect of going from 1 ppm to 100 ppm is much greater than going from 200 ppm to 1000 ppm. The change in flux from 1 to 100ppm is 24 W/m^2, while going from 200 to 1000 ppm is 9 W/m^2. If it takes the ~32 W/m^2 of flux associated with a CO_2 of 380ppm to cause 1.0°C of the total GHE, then an increase of 10% will only produce ~0.1 °C of actual temperature change.

Why the flux changes like this is not a simple discussion. I will go over infrared absorption in depth later in the Scientific Content section. Many of the more hotly contested topics will be covered there. I tried many ways to simplify it, but in the end I chose to explain the results in the main chapter and just cover the science in that section. I might one day put a more complex analogy I call the Stained Glass Greenhouse up on the website. It was once here in this chapter, but for now I am letting it stay out of the discussion.

If the Theory of Global Warming was correct, the Eemian would not have ended when the 65N insolation anomaly went negative 120,000 years ago. There was almost no change in the TSI over that period of time. There was no change in atmosphere at that time. The only thing that changed was the insolation at 65N. It dropped dramatically over the course of 15,000 years. Elevated

CO_2, warm oceans, high humidity did nothing to prevent the Earth from cooling.

Since the atmospheric and ocean conditions did nothing to prevent the Earth from cooling in the past, there is no reason to predict that insignificantly tiny differences today from the Eemian will make a difference in the future of the Earth's climate. What will make a difference is the amount of energy that 65N will get in the future. That has been decreasing now for 9,000 years and the anomaly has been negative for the past 3,000 years. There is a reason that glaciers have been forming over that period of time. It has nothing to do with changes in the atmosphere and everything to do with changes in summer insolation.

There is one other way to make CO_2 appear important to the Earth's climate. If it is assumed that a particular change in temperature happened because of a change in CO_2 level, then the result will be that the Earth is very sensitive to changes in the level of CO_2. Of course to do that requires carefully selected periods of time that fit with the Theory of Global Warming.

This leads to the topic of Climate Sensitivity, which states that a change in the radiative flux (referred to as forcing by most warmists, but it is really just a measure of radiative flux) will cause a specific change in temperature. So if the Earth was warmer in 2005 than it was in 1990 and the CO_2 level changed from 355 to 380 ppm in that period of time. Then the climate sensitivity would be.

$$\lambda = \Delta T / \Delta F$$
$$\lambda = \text{climate sensitivity (W/m}^2 / K)$$
$$\Delta F = \text{change in forcing, } \Delta T = \text{change in temperature}$$

$$\lambda = 0.27 \,°C / 0.48 \text{ W/m}^2$$

$$\lambda = 0.56 \text{ K} / \text{W/m}^2 \text{ (2.1 °C/doubling)}$$

That is usually converted into terms of temperature change for a doubling of CO_2. The conversion is to multiply the climate sensitivity by the 3.7 W/m^2 associated with a doubling of CO_2. In this case the result will be to predict that a doubling of CO_2 will cause the temperature of the Earth to increase by 6.3 °C. So it is very possible to reach a conclusion based on forcing that increasing the amount of CO_2 will cause enormous warming.

Of course if I picked 1998 to the year 2010 the result would be very, very different.

$$\lambda = -0.043 \,°C / 0.39 \text{ W/m}^2$$

$$\lambda = -0.11 \text{ K} / \text{W/m}^2 \text{ (-0.4 °C/doubling)}$$

In this case the increase in CO_2 causes cooling. That is why this method only works over periods when the temperature is warming. Applying this method to the start of interglacials is a very popular scientific pastime for warmists. Applying it to the end of interglacials is not very popular as the result is always a negative climate sensitivity.

The warmist 'consensus' is that doubling the amount of CO_2 in the atmosphere will cause the temperature of the Earth to increase by 3.0 °C. That corresponds to a climate sensitivity of 0.81 $W/m^2 / K$. I prefer to use the less biased version of climate sensitivity. Here is a quick conversion table for climate sensitivities.

Δ °C/(2x CO2)	1.6	2.0	2.5	3.0	3.5	4.0	4.2
λ °C/(W/m2)	0.43	0.54	0.67	0.81	0.94	1.08	1.13

Illustration 141: CO_2 doubling values on top and actual climate sensitivity values are below. This covers the range of possible climate sensitivities used by the IPCC and most warmists.

As a concept I actually like climate sensitivity, but each year shows the limitations. The temperature of the Earth changes 4 °C with no real change in forcing, even an opposite change in forcing as the TSI is highest in January while the Earth is the coldest.

However it did lead me to what I call the Natural Climate Sensitivity which I define it as the total energy transferred to the atmosphere and the total GHE.

$$\lambda = \text{Total GHE} / \text{Total Energy}$$

$$\lambda = 33 \text{ °C} / 120 \text{ W/m}^2$$

$$\lambda = 0.28 \text{ K} / \text{W/m}^2$$

The basis for this is the same as how a furnace is rated for warming a house. So much energy will keep this much space warm, so many kW's (or BTU's) will only warm up so much space. For the Earth it takes a lot of energy to keep the Earth warm. This allows for an easy measure of how much of the total GHE is caused by a particular energy flow. Evaporation is responsible for 80 $W/m^{2 \text{ of}}$ energy to the atmosphere. That makes it responsible for:

$$\text{Evaporation GHE} = \lambda * \text{Energy Evaporation}$$

$$\text{Evaporation GHE} = 0.28 \text{ K} / \text{W/m}^2 * 80 \text{ W/m}^2$$

$$\text{Evaporation GHE} = 22.4 \text{ K}$$

So the transfer of energy from the surface to the atmosphere by evaporation is responsible for 22.4K of the GHE. This is a different approach, but it is a very practical approach where each contribution to the GHE can be shown by its significance.

The Theory of Global Warming depends on changes in forcing to cause drastic changes. The problem is that in all other applications of heat transfer, the type of changes the Theory of Global Warming uses does not cause significant changes in the amount of heat that is transferred.

Radiative heat transfer is not new or special. Scientists and engineers have been dealing with it on a practical level for a very long time now. The idea that a small change in the radiative flux is going to cause significant changes in the Earth's temperature is just not a good conclusion based on the science of heat transfer. In most practical applications of heat transfer radiative is only the dominant factor under a few very special conditions.

1. Very high temperatures with a large temperature difference between objects.
   (example is a campfire on a cold night)

2. Under very low pressure where the other forms of heat transfer cannot occur.
   (example is in space where there is no atmosphere)

Neither of those conditions applies to the Earth's surface and the atmosphere. As a result the effects of radiative heat transfer should be very comparable to that of convection. The results of greenhouse gas absorption and convection show that they are in fact very comparable in size to the amount of energy that is transferred.

Understanding the Effect of Doubling the Concentration of CO_2

There are many real world applications where it is critical to know how much infrared energy will be transmitted. One of the more interesting applications is in military targeting systems. In this case IR beams will be used to 'paint' a target for precision bombs. It is absolutely critical to their effectiveness to know if the energy that is transmitted will reach the target. If the atmosphere absorbed the energy from the IR beam, it would be worthless. Infrared is popular because it cannot be seen by the human eye and there are wavelengths that are not absorbed by the atmosphere.

Illustration 142 shows the transmission for the atmosphere between 6-18 micron wavelength. This is the primary spectrum that the Earth's surface transmits IR energy at. More importantly it shows the transmission of IR through 1km of atmosphere and from the surface to the top of the atmosphere, a distance it puts at 100km. This single chart by itself is enough to disprove the theory of global warming. The difference between the two lines is the difference between 1km of atmosphere and 100km of atmosphere for IR transmission. For CO_2 in the atmosphere, it is the region between 13-16 micron that matter.

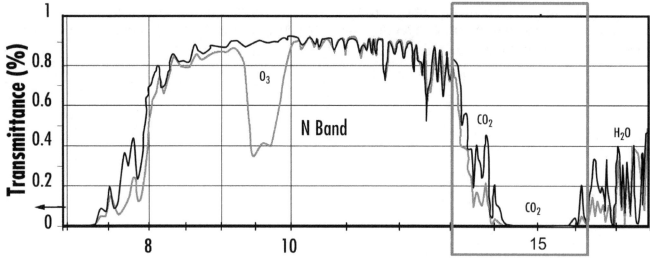

Illustration 142: (High) 1km Horizontal Transmittance (Low) Transmittance from Sea level to 100km Vertical. I added 15 to the x-axis as that is the peak CO_2 absorption. Notice how little difference there is between 1km of atmosphere and 100 km of atmosphere for IR transmission in the CO_2 spectrum.

The difference between the Red and Black lines between 13-16 micron is the total amount of energy that CO_2 would be able to absorb at ~80 times the current concentration. That is the difference between 1km of atmosphere and 100km of atmosphere. I even give the benefit of the doubt since the atmosphere drops of significantly at 80km. When the amount of energy from the source is absorbed at > 99%, then for all intents and purposes it is not possible to absorb more energy. That there is no significant difference in the total transmission between the lines in the range that CO_2 absorbs energy is absolute evidence that increasing the concentration of CO_2 will not significantly increase the amount of energy that is absorbed by the atmosphere.

236

The blackbody spectrum for an object with the Earth's temperature is shown in Illustration 143. The primary portion that CO_2 absorbs is the 13-16 micron. Of the energy that the Earth transmits, that is the ONLY portion that CO_2 can absorb. If CO_2 absorbed 100% of that energy (which it almost does) transmitted in that band, it would absorb ~3% of the total IR from the Earth's surface. Doubling the concentration of CO_2 cannot increase the total amount of energy that is absorbed.

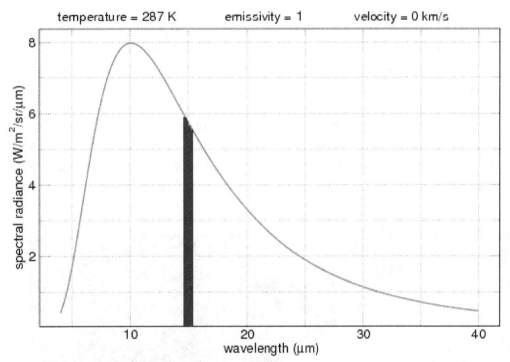

Illustration 143: The band that CO_2 can absorb. Courtesy of SpectralCalc.com

Two things happen as the concentration of CO_2 doubles. The first is that the distance it takes to absorb the same amount of energy is cut in half. So if it took 2 km to absorb 100% of the energy it would now take 1 km to absorb that amount of energy.

The second thing is broadening. That is the edges of energy expand a very slight amount. That is the basis for the increasing energy of CO_2 as the concentration increases. The actual change in the energy absorbed has almost no change though because almost all (~99%) of the energy has already been absorbed and converted into heat.

At this point every warmist that is following along is jumping up and down saying that it is the re-transmission that matters. Wrong!!! I will have to use examples to explain it though. There are two and I will use both of them even though they are related.

The first is a very cloudy day. On such a cloudy day, even at noon, with the Sun high in the sky, the Sun is not visible. 100% of the visible spectrum has been absorbed by the clouds and none of the visible light from the Sun directly reaches the ground. If it did, the Sun would be visible. What is seen is the light being re-transmitted by the clouds.

The maximum amount of energy that can be blocked is the total amount of energy from the Sun at the location in question. It doesn't matter that light is being re-transmitted, the total amount of energy is the same. The thicker the clouds are, the darker the day is because the re-transmitted energy will always be less after each re-transmission. Each layer of clouds ends up decreasing the available light.

Once again energy is energy. It doesn't matter how a gas molecule gets the added energy (collision or absorption), it will either warm up or transmit a photon. Some molecules warm up, some re-transmit a photon. As the clouds get thicker the number of photons will decrease and the day will get darker.

No matter how thick the clouds get, the maximum amount of energy that is being blocked is the amount of energy from the Sun. As the clouds get thicker the percentage of sunlight that is blocked will approach 100%. Once the clouds block 100% of the sunlight, additional clouds will not block more energy, but the photons that get re-transmitted will decrease as the clouds get thicker.

Illustration 144: Supercells can prevent almost all sunlight from reaching the Earth's surface. CO_2 acts in an identical fashion at blocking the 14-16 micron wavelength from reaching space. Martin Lisius/StormStock

The conservation of energy requires that the energy in is equal to the energy out. If 100% of energy is converted to heat, then increasing the concentration will absorb only additional energy which is caused by spectral broadening. Spectral broadening will increase energy absorption by a minute fraction. Re-transmission is not the creation of additional energy.

That is the more accurate situation for CO_2 in the atmosphere today. It only takes a few km of atmosphere to absorb almost all of the energy that it can absorb. Sure the atmosphere is always transmitting at the proper wavelength, but any gas with that concentration of CO_2 would transmit in the same wavelength because it is the correct temperature to transmit at that wavelength.

The other example I want to use is fog. This is a more accurate representation of CO_2 on the IR energy from the Earth's surface. On a foggy day there is limited visibility. Objects in the distance cannot be distinguished because 100% of the transmitted light is absorbed by the fog. Only light from nearby objects can be directly seen. Doubling the thickness of the fog only has the effect of reducing the visibility by half. That is precisely the effect that doubling the concentration of CO_2 will have on the IR energy from the Earth's surface. The distance it takes to absorb the 3% of energy that can be absorbed will be cut in half. The energy absorbed over a particular distance will increase, but there is almost no change to the total amount of energy that is absorbed.

Illustration 145: Visibility is a measure of how much light is being absorbed by water in the air. Doubling the fog cuts in half the distance that light will travel before being absorbed. Courtesy of GPB.org.

Gases transmit IR energy based solely on their temperature. Cold gases transmit very little IR energy because only a very tiny fraction of gases reach the energy level required to transmit a photon. That changes as the temperature increases. The Earth's atmosphere transmits IR based on it's temperature, exactly as predicted by gas kinetics. The only change that takes place based on gas concentration is the specific wavelength. It is that combination of factors that allows the temperature and composition of other atmospheres in the solar system to be determined from the Earth. Changing the concentration of gases alters the wavelength, but not the total amount of energy that is transmitted by gases.

I would like to end by comparing the spectral modeling of doubling the concentration of CO_2 to doubling the distance by using the HITRAN2008 model available at the SpectralCalc[153]. The starting baseline transmission will have a CO_2 concentration of 300 ppm and a distance of 500m. The transmission for these conditions is 0.8775.

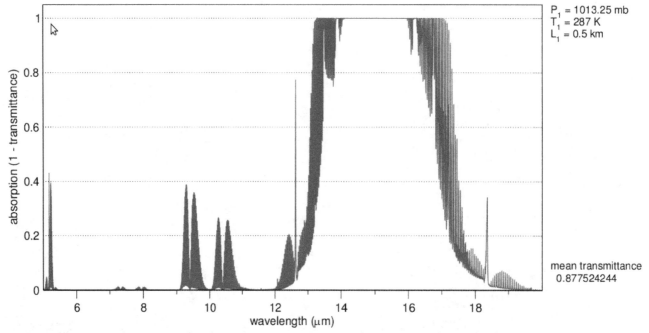

Illustration 146: 300 ppm CO_2 and 0.5km of length.

In other words, these conditions absorb 12.25% of the energy between 5-20 micron with most of that absorption taking place in that key 13-16 micron range. This shows absorption above 16 micron, but I will ignore that because in the real world because absorption above 16 micron is dominated by water vapor.

Doubling the concentration of CO_2 should be identical to doubling the distance. In both cases this will be indicated by a decrease in the transmission. The change in transmission will be small because the wavelengths that CO_2 can absorb are limited.

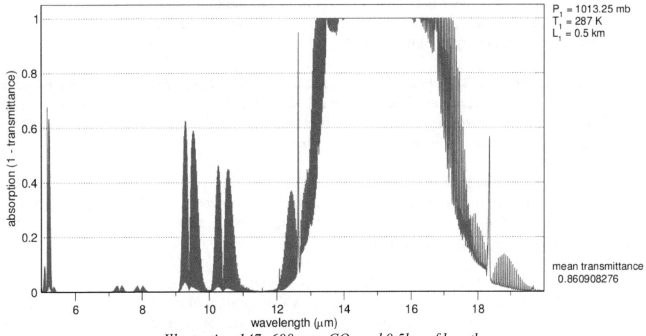

Illustration 147: 600 ppm CO_2 and 0.5km of length.

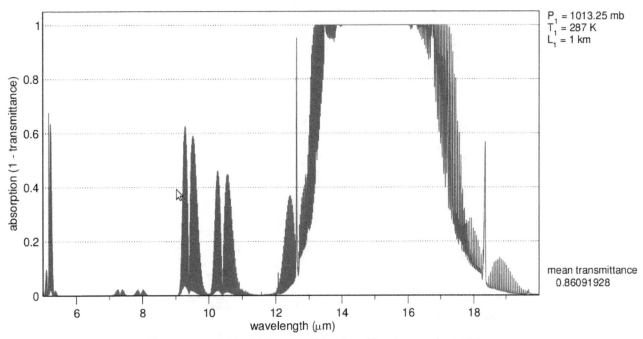

Illustration 148: 300 ppm CO_2, but 1km instead of 500m.

In both situations, the amount of energy that is absorbed increased from 12.25% to 13.91%. So while the amount of energy absorbed did increase, it shows that doubling the concentration has the same effect as double the distance.

Since the troposphere is ~12km long and the concentration of CO_2 is nearly constant for the length of the troposphere, there is almost no real change in the total energy absorbed by 12km of CO_2. Even if CO_2 was the only factor, the only real effect would be that the atmosphere would absorb that quantity of energy in 6km instead of 12km.

When this information is applied to the overall energy balance, the change in energy absorbed across 12km of atmosphere going from 300ppm to 600 ppm is ~0.56 W/m^2. This also assumes that CO_2 is the only gas absorbing energy, which is also not the situation as spectrum broadening past 16 micron is dominated by water vapor absorption.

Doubling the concentration of CO_2 from 300ppm to 600 ppm will add ~0.56 W/m^2 of additional energy to 12km of troposphere. That is an increase from ~120 W/m^2 to ~120.5 W/m^2 or a total maximum change of 0.4%. One day, we might even one day have the technology to accurately measure that small of a difference.

Since it takes 120 W/m^2 to generate a GHE of 33°C, that change in energy would be enough to increase the GHE by 0.1 °C. That is ~30 times less than predicted by the warmist models. So there is a difference, but it is not one that will cause any measurable change to the Earth's climate. Even if it did, what I cover next will explain why it doesn't matter.

Chapter 13 The Limits of Temperature Change

Scientists must always consider the fact that their conclusions could be wrong. Despite everything I have found up to this point in the book I still did not dismiss the idea that the increased CO_2 levels could have a bigger impact than what the energy balance indicated. Keeping an open mind is really the key to being a good scientist or engineer.

In my professional life as an R&D engineer in the semiconductor industry there have been some very unusual solutions to problems that I would never have expected to work. Strange new solutions to problems do happen in the real world. Keeping an open mind to new ideas is absolutely critical to being successful in the highly competitive semiconductor industry.

In that line of work there is no peer-review system to determine if an idea is valid or not. It is simply tested on the real world product and if it works, then the idea was a good one. If it doesn't work then the idea isn't used. One of the key goals in real world R&D is to learn as much as quickly as possible and apply it in the most cost effective way.

There is a reason that phones today are as powerful as desktops were 5 years ago. The company I work for and the others out there have been very successful at finding new and better ways to do things. So successful has the semiconductor industry been over the past 40 years at constantly improving technology that many people expect it to always be that way. I have seen so many seemingly impossible barriers surpassed in my 12 years as an R&D engineer that it almost makes me believe that it will continue that way for the foreseeable future even though I know that some really difficult barriers are looming large.

I tell you this because I want you to understand why my approach is more typical of research and development in industry than in a university. Getting published is the end goal of a professor. Getting new technology to work is mine. If the new knowledge isn't applied then the idea was not a success. It might not have been a failure and might work in the future, but it was not a success.

My goal right now is to alter the entire debate about global warming. What I have presented so far is the most coherent review of the information available that I have found, but it doesn't really change anything. Like most really interesting ideas I wasn't really looking for it when I found it.

When I was looking at was the 65N insolation peak that took place 105,000 years ago. It didn't make sense to me. Many other times over the past million years there have been similar increases in insolation that didn't trigger an interglacial. This was simply one more such instance. This one particularly got my attention because the anomaly was higher 105,000 years ago than the Holocene increase in insolation.

Vostok Temperature Anomaly

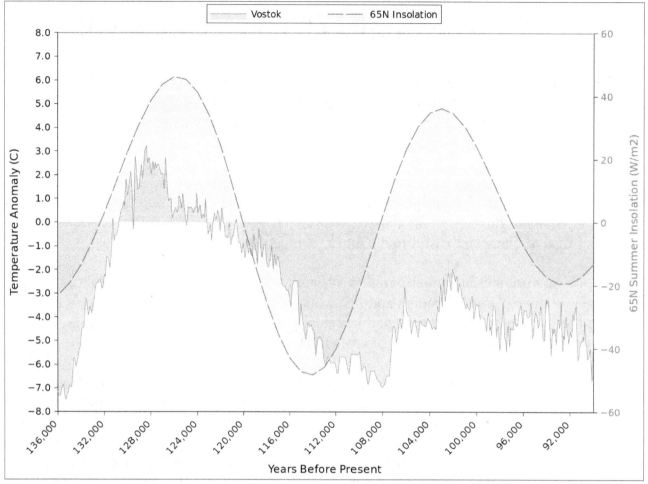

Illustration 149: Two peaks in insolation resulted in only one interglacial.

The temperature did increase a little, but not very much in comparison to what it did in the Eemian. I didn't really feel bad as this is something that scientists used to look at before CO_2 became the boogeyman. Now they only look at how CO_2 is the cause of everything. Of course CO_2 even made less sense because the level was still high when the insolation started to increase. The temperature increased slightly with insolation, but the correlation to CO_2 was non-existent.

Prior to the Eemian, the CO_2 was even lower than it was 108,000 years ago when insolation increased. So if anything the theory of global warming would say that the Earth should have warmed up even more than the Eemian 105,000 years ago, but it didn't really warm up at all.

The drop in insolation was large at the end of the Eemian, but the Earth moves slowly in the long-term. That is when I realized the Earth wasn't as cold when the insolation started to increase 108,000 YBP as it had been when the insolation started to increase at the beginning of the Eemian. A warmer Earth loses energy at a higher rate. This means that less energy was available to cause warming.

Vostok Temperature and CO2

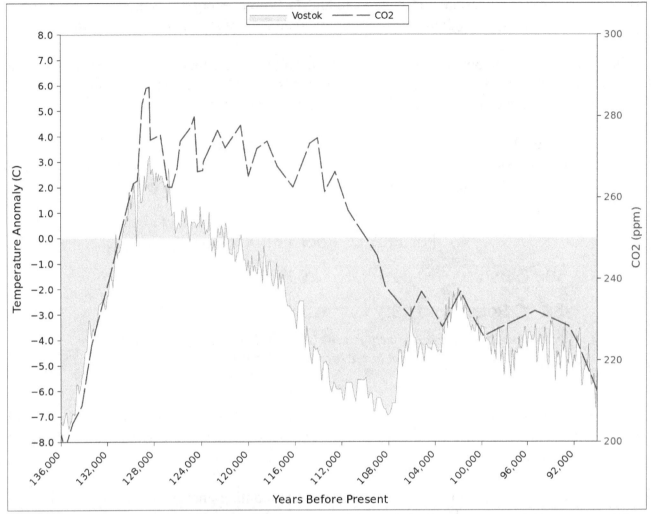

Illustration 150: The CO₂ level and increasing insolation 105,000 years ago should have caused the Earth to warm up, but it didn't.

This is the one place where the Radiative Heat Transfer really, really matters. The amount of energy that the Earth is losing is dependent on the temperature of the Earth. 105,000 years ago the Earth had cooled down a bit, but it was still warmer than it was 20,000 years ago when the Holocene started. That meant that the gap between the incoming energy from the Sun and the outgoing energy was not large so there was only a small increase in temperature.

Now to explain that in a way that makes sense; the Earth is not in equilibrium.. It might look that way in the short-term, but the Earth changes temperature slowly in the long-term. Let's say that the NH is, on average, losing energy at the rate of 200 W/m². If the Earth was in a glacial like it was 20,000 years ago that value was closer to 150 W/m². When the 65N summer energy started to increase 20,000 years ago the NH was getting a large excess of energy because the temperature was very low. When there is a lot of extra energy things warm up quickly.

105,000 years ago in the NH the temperature was not nearly as low. It was still cooling down. If the NH energy loss was 180 W/m², the extra energy was there, but the difference between the energy being lost to space and the change in the incoming energy was not that great, so the temperature rose slowly enough that the 65N dropped again before the NH warmed up.

Time for a simple analogy; if I have two pots of boiling water that are each 100 °C and I stick one into a room that is 50 °C and one into a room that is -50 °C, which one will cool down faster? The more important question is why will one cool down more quickly.

If I simply use RHT for this situation it will be more comparable to the Earth losing energy to space.   In the two situations the difference in radiative flux will determine how quickly the temperature changes. Remember that flux is simply the infra-red 'heat' of an object.

	Pot Flux	Room Flux	Energy Transfer
50 °C Room	1099	618	481
-50 °C Room	1099	148	951

Illustration 151: All flux and heat transfers are in W/m². Energy is flowing from the pot to the room in both cases.

The pot in the cooler room loses energy at almost twice the rate and so it cools down much more quickly that the pot in the warmer room. This should fit well with the observations that people make on a daily basis.

In the case of the NH, both the temperature and the incoming energy are changing. This leads to two distinct possibilities. There is the situation when the NH is cold which helps limit further cooling, but it provides the opportunity for it to warm up. The reverse situation is when the NH is warm (like it is today) which makes it difficult to warm it up further, but makes it very easy for it to cool down. The shapes of the slopes are odd, but that it due to the temperature dependence.

The warmer the Earth gets, the more energy it takes to maintain that temperature. If the Earth was a mere 5 °C warmer, it would lose an additional 18 W/m² of energy. For as long as the Earth stayed at that higher temperature, the Earth would continue to lose energy at the higher rate.

During the last glacial period large portions of the NH were ~10-12 °C[154] colder than they are today. Those regions are losing 30 W/m² more energy today than they did during the glacial. That lower rate of energy loss prevented the Earth from getting even colder when it could have continued to cool. It also means that the situation now favors cooling instead of warming.

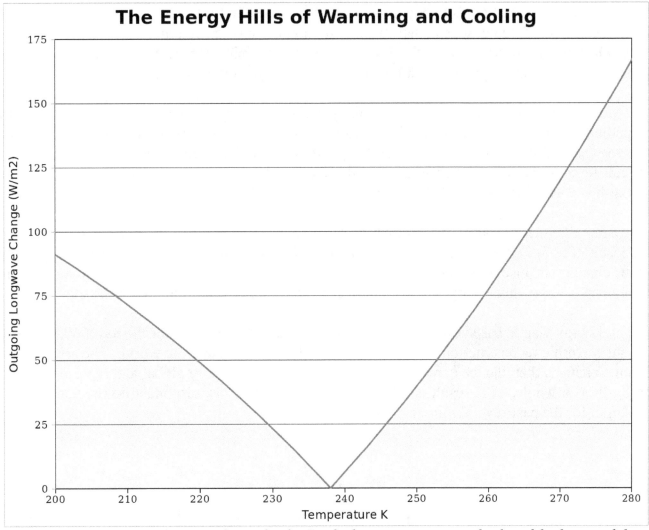

Illustration 152: A novel way to show why the Earth changes temperature back and forth around the central temperature. Any anomaly from the equilibrium temperature is like a rock being rolled up a hill. It is much easier for the rock to go downhill than it is to go uphill.

This is why the NH has been cooling for the past 3,000 years. Since the peak energy of the Holocene took place 9,000 years ago, the 65N insolation has dropped 40 W/m². The only possible outcome of this is for the NH to cool down. It is still losing energy at the higher rate, but getting energy at the lower rate. Since the overall change in energy takes place over thousands of years, it takes that amount of time for the temperature to change as well.

In the end there are two special circumstances that arise as a consequence of the dynamic changes to temperature and insolation.

1. Cold Earth with increasing insolation.
2. Warm Earth with decreasing insolation.

When those situations occur together, the Earth can switch to the other state. The temperature of the Earth is the critical trigger that ties the insolation from the Sun to the glacial/interglacial cycle.

In situations where the Earth's temperature is more neutral, the Earth responds more slowly. The situations where it would appear that insolation should cause an interglacial, but did not, took place when the Earth was not at the lower temperature limit. Hence the NH was still losing energy at a higher rate so the Earth did not experience significant warming.

On a playground swing-set the bigger the push, the higher you will go. Very cold or warm temperatures result in big pushes for large temperature changes to take place, but the same effect puts very natural limits on how much the temperature of the Earth can change. It would be very difficult to cause additional warming when the Earth is hot or to cause additional cooling when the Earth is cold. The hill simply becomes too steep to climb.

I define the Energy Gap as the anomaly in the amount of energy that the Earth is losing energy as a result of above or below average temperatures. When the absolute value of the energy gap is small, changes in insolation cause small changes in temperature. When the magnitude of the energy gap is large, then changes in insolation can cause large changes in the Earth's climate.

The Energy Gap is based on the average temperature of the Earth[155] for the past 500,000 years using the benthic reconstruction[156] from 57 different locations around the world. The result is less extreme changes than the ice cores provide, but the coverage is more global and representative of the Earth as a whole. The result is a basis for why the Earth has responded to the variations in insolation for the past several million years.

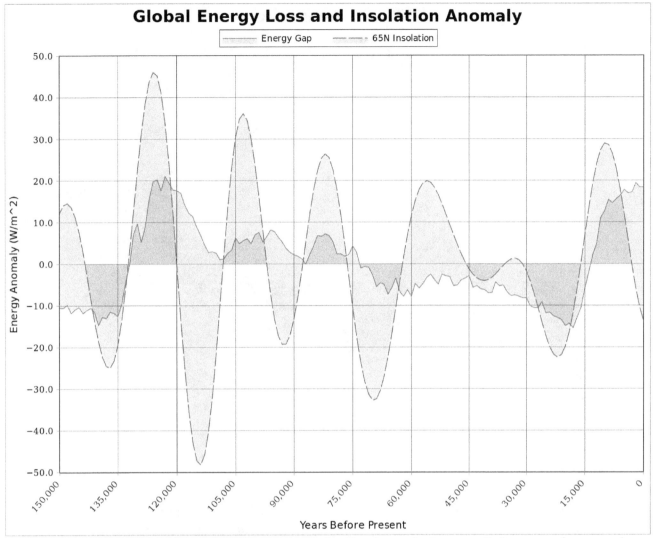

Illustration 153: 135,000 and 20,000 years ago the Earth was losing very little energy because the temperature was cold. When the energy from the Sun increased the Earth experienced dramatic, but limited warming. 115,000 years ago and today the Earth is warm and losing excessive energy to space. 115,000 years ago the Earth experienced significant cooling.

This is the reason why the amount of warming that the Earth can experience is limited. There are physical limitations on how much warming can occur. The warmer an object is, the more energy it loses. The Earth is no exception and the amount of CO_2 in the atmosphere cannot change that.

Since 1979, science has had the capability to measure the amount of energy that is leaving the Earth. It is called the out-going longwave radiation (OLR). This infra-red energy is how the Earth loses energy to space. It is possible to see how the OLR has responded to the temperature change that has taken place over the past 30 years. It should be no surprise that the amount of energy escaping the Earth has increased[157] with the temperature. That is the most basic rule of radiative energy.

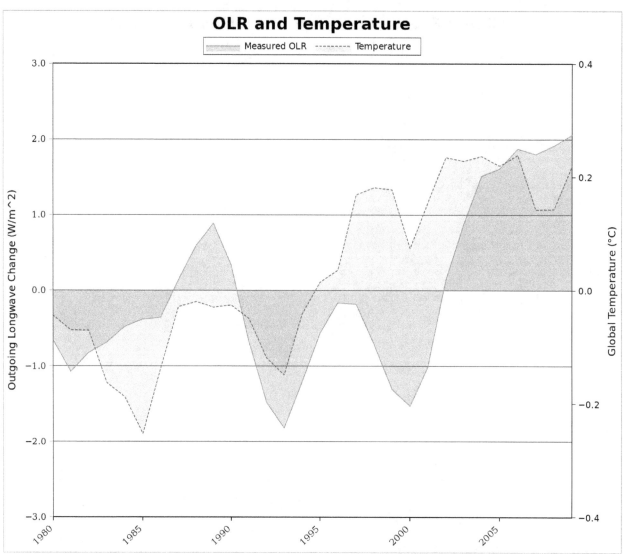

Illustration 154: Since 2002 the Earth has been losing an additional 2 W/m² of energy to space. The only possible outcome is cooling in the future.

A 0.2 °C change in temperature resulted in a 2 W/m² increase in the amount of energy the Earth is losing. That energy is simply gone forever. Any additional increase in temperature will result in even more energy being lost to space. There is no way that can be avoided.

If I compare the increase in OLR for the IPCC projection that the Earth will warm up 3 °C for a doubling of CO_2 it is possible to plot out the projected increase in OLR that will result as the level of CO_2 in the atmosphere increases. Remember that the 'forcing' for a doubling of CO_2 is 3.7 W/m². I am even going to base that 3.7 W/m² as being achieved at 600 ppm which is a doubling from 300 ppm, but base the increase of OLR as from the modern day. That gives the theory of global warming a 100 year head start. My projection for a future CO_2 level at 600 ppm in 2100 is pretty much in line with predictions[158].

The result is that the 'forcing' of CO_2 is 3.7 W/m². The Earth would be warmer by 3 °C according to the Theory of Global Warming, but the problem is the Earth would then be losing 7.5

W/m² more energy than it does today. The Earth would be losing more than twice the energy to space than the 'forcing' effect of CO_2. That ratio increases the more the Earth warms up. Warming up the Earth will always ensure that it will cool down in the future.

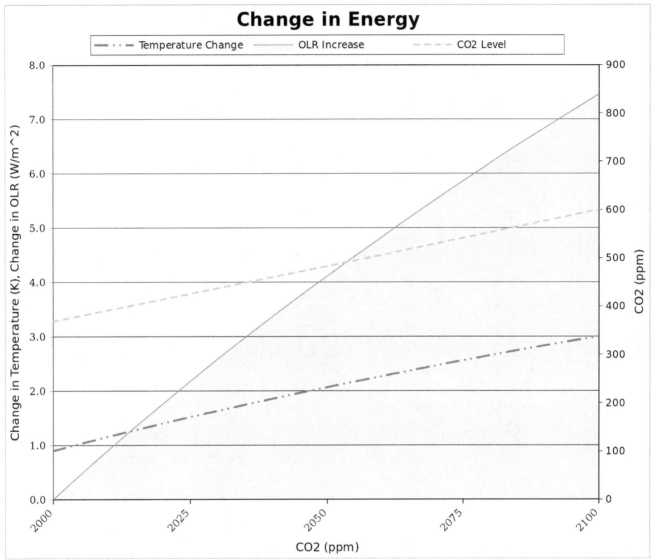

Illustration 155: The warmer the Earth gets the faster it loses energy. When the Earth is as warm as it is today, it is very difficult to cause additional increases in temperature.

This also highlights one area where temperature anomaly is not sufficient. When it comes to the amount of energy that is radiated by the Earth, it is the actual temperature that matters. That is because the OLR is determined by the actual temperature of the Earth and not the temperature anomaly. This is easily shown because the measured OLR tracks with the actual temperature of the Earth.

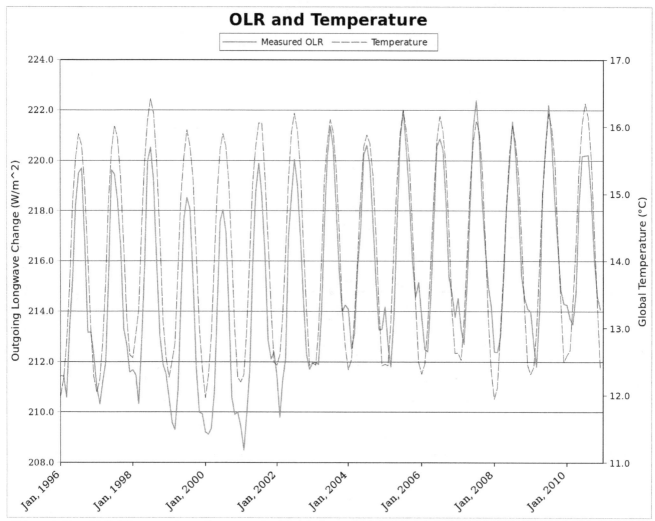

Illustration 156: The amount of energy the Earth loses is in sync with the Earth's seasons. The Earth's temperature and not the amount of energy available from the Sun determine the amount of energy that the Earth is losing.

Each NH summer, the amount of energy escaping the Earth increases with temperature, even though the Earth is getting 6% less energy than it does during the NH winter. That both the winter and the summer have been losing extra energy is absolute proof that the Earth is warmer, but it also proves that the limits to temperature change are in place. The warmer the Earth gets, the faster it loses energy. That is the cause of the natural limits of temperature change for the Earth.

This can be used to show the dependence of OLR on the Earth's temperature. This cannot be

252

shown based on temperature anomaly because a warmer than average January is still much colder than July. So a warm January will lose more energy than a cold one, but they will always lose less energy than even an abnormally cold July. That is the most important reason why temperature anomaly has no true value in monitoring the Earth's climate.

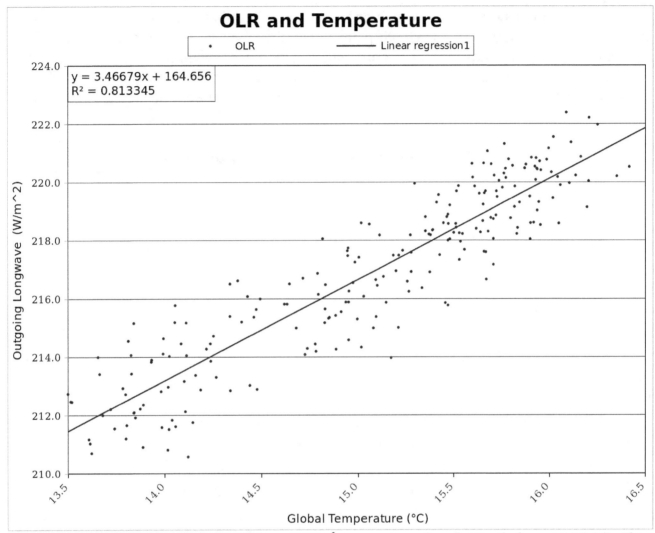

Illustration 157: The Earth at 16.0 °C loses 7 W/m² more energy than the Earth does at 14.0 °C. This is the main mechanism that regulates the temperature of the Earth. The trend is ~3.5 W/m2 per °C.

This takes place on all of the climate time scales. Every day the Earth starts the morning losing less energy than it will in the afternoon. This is why the evening cools down more quickly than the middle of the night.

Every year the NH starts off losing less energy than it will in the summer. This limits how hot the summers get because higher temperatures cause more energy to be lost to space. It also limits how cold the winters get because the rate of energy loss decreases.

It also takes place on the time scale of hundreds of years. This is what happened during the Little Ice Age that took place 250 years ago. The Earth got cold and this slowed the rate of energy loss. This resulted in a negative energy imbalance that caused the Earth to start warming up. It will keep warming until the Earth is has swung too far and then it will start cooling again. The Earth is far too dynamic a system to reach a true equilibrium, so instead it swings back and forth around the current long-term equilibrium temperature.

That is the real answer to the question about why the Earth has warmed up over the past 150 years. This is the cause of the sawtooth pattern that shows up in every measure of the Earth's temperature. Every day, season, year, century and millennium show limited warming and cooling. Never has the temperature of the Earth changed without bounds. The past 8,000 years from the EPICA ice core show constant and significant swings in temperature, but they are all limited in scale.

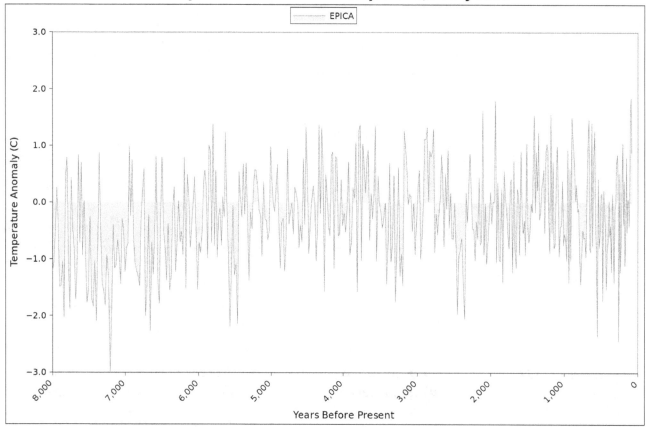

Illustration 158: While there is no way to show that changes in the rate of OLR caused the past temperature swings, the laws of physics are that warmer bodies lose energy at a faster rate than cool objects. OLR explains the behavior that CO2 levels do not.

When the Earth cools in a glacial period, the amount of energy that the Earth loses to space drops by huge amounts. A 10 °C drop like the kind experienced in the Last Glacial Maximum reduces the amount of energy lost to space by a minimum of 32.5 W/m². That is a 15% decrease in energy loss. That is why the Earth tends to swing back and forth between temperatures. As the amount of energy lost decreases, the rate of the cooling decreases until eventually, the Earth stops cooling. This sets the Earth up for rapid warming later on. In a deep glacial period the rate of

254

energy loss is at the minimum. When insolation increases, the Earth warms up quickly. When the Earth is warm, it is impossible to cause rapid warming.

This is why the reconstructed temperature of the Earth for the past 5 million years shows a very limited range of temperatures. While the variability has increased primarily as a result of increasingly cold and long glacial periods, the upper limit of the Earth's temperature has changed very little.

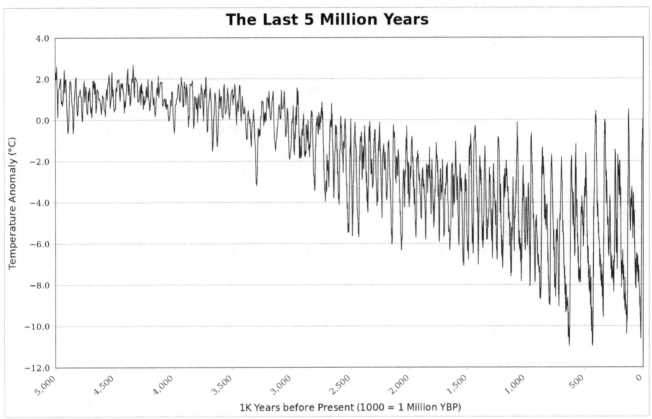

Illustration 159: The Earth's temperature is always varying from one extreme to the other.
(Raymo, 2005)

It has an upper limit to the temperature and a lower limit. The initial changes take place quickly then slow down as the OLR brakes are applied. Warmer temperatures make it easier for the Earth to cool quickly and cooler temperatures allow the Earth to warm up quickly. Additional warming when the Earth is already warm would require a permanent increase in the amount of energy it gets from the sun.

OLR accounts for about 68% of the Earth's energy balance. It is also directly dependent on the actual temperature of the Earth. As the Earth's temperature changes, the amount of energy the Earth is losing adjusts to the new temperature. That is why the Earth has stayed within the +/- 6 °C of the average temperature for the past 5 million years.

OLR is the single most important regulator of climate.

A distant second is the Earth's oceans. They act as a secondary storage for energy when the Earth cools and as an air conditioner when the Earth warms. The effects are much smaller than those of OLR. Increasing temperature in the oceans results in increasing evaporation. That evaporation causes cooling of the oceans. When the oceans cool there is less evaporation which slows the cooling of the oceans. This secondary cooling system is seen in the El Nino and La Nina events. Those changes take place every few years. There is literature that indicates when the Earth is warmer, the El Nino is more constant[159]. The same paper also points out that the warm phase has been in decline for the past 1,200 years which is in line with the decreasing 65N insolation. The temperature response to decreasing energy is evident everywhere, even in the oceans that help regulate that same temperature.

The Earth has generally been stable for billions of years. There are the extremes though. There are a couple of times where it appears that the Earth entered what is called a "snowball" state where it was almost all (if not completely) covered by ice sheets[160]. The most recent of these happened ~570 million years ago. In that state the amount of energy lost by the Earth to space is very low. If the Earth was 20 °C colder in that state, it would on average lose 75 W/m^2 less energy than it does today. Such a state is inherently unstable with the amount of energy that the Earth gets from the Sun and would rapidly warm up when the right conditions developed.

There is no scientist that will say that increasing temperature should not increase the OLR. They might argue that CO_2 will inhibit it at some point, but the measurements do not support that theory at this time because the OLR has increased precisely as it should with the Earth's actual temperature. That means that while the Earth is the warmest it has been in a few hundred years, it is also losing energy at the highest rate that it has in that same period of time.

Especially since the amount of energy the Earth is losing has increased in the last 20 years. In the 1980's the measured OLR was 214 W/m^2. For the past 8 years that average has been over 216 W/m^2. That 2 W/m^2 change seems small, but that is a real and measured 0.6% change to the whole balance of the Earth's energy. It is also many times greater than the effect of doubling the concentration of CO_2 in the atmosphere[161].

To put that into human terms, each hour the Earth is losing an *__extra__* 3,670,000,000,000,000,000 KW-hours of energy to space. Or, every second the Earth is now losing enough extra energy to power everything that mankind uses for more than 50 years. That is not theory, that is measured observation[162].

That is precisely why the Earth stays within certain temperature limits. Greenhouse gases will always be limited to their effect. So will evaporation, volcanoes, ice ages, glacials, interglacials and everything but significant changes to the sun.

There is no limit to the effect that OLR has. The more the Earth warms, the more that OLR tries to cool the Earth. The colder the Earth gets, the more OLR prepares the Earth to warm up.

OLR is the safety valve of the Earth's temperature. It naturally opens up more when the Earth warms and closes when the Earth cools. That is why there is no reason to worry about global warming. That is why there is every reason to worry about the next glacial.

The Earth is currently at a real temperature anomaly of ~5 °C. That puts the current energy gap at +18 W/m^2. Causing the Earth to warm up would require that a source of energy be available to make up that energy gap. Instead of excess energy, there is a deficiency for the NH that currently sits at -14 W/m^2. The Earth is currently in the same state today that it was 119,000 years ago when the Eemian started to experience significant cooling.

That is the current state of the Earth's climate. I told you that I would "show you the data." Well, there it is. The situation of the Earth is very clear.

Summer is over.

Fall is fading.

Winter is coming.

Warning:

Scientific Content!!!

The Energy Gap:

There is really no way to describe how important the development of the energy gap idea is to understanding the Earth's climate. Science has in many ways taken a break when it comes to developing new ideas about the climate. The primary reason is that everything has to fit the Theory of Global Warming. The idea of the energy gap is anathema to that theory. Hence it could not be developed by someone trying to explain the Earth's climate in the narrow confines of the cult of global warming.

It explains everything from why the Earth has been warming for the past century, but it also explains why it cooled down prior to that. The Medieval Warming Period caused the Little Ice Age. The LIA caused the modern warming period which I like to call the Al Gore Warming (AGW). It seems a better fit than Anthropogenic Global Warming.

It also explains why temperatures change quickly and then stabilize before changing again. This is exactly what the Earth is doing right now. It warmed up until the 1998-2002 period and then it stopped warming. At that point the OLR caught up with the temperature and the Earth stopped warming. Now the Earth has a higher OLR than it has ever had before (at least as measured by our instruments). For the past 5 years scientists have been confused about the energy that should be somewhere, but isn't. I have a pretty good idea. OLR picked up because the Earth is warm and that energy is gone forever. The Earth will keep losing energy at the higher rate until it cools down again.

The real question is when will it happen and how cold will it get. I don't have that answer, but my best WAG is that the longer it takes to cool, the more the temperature will drop. It is even possible that there will be more warming before the temperature eventually drops. The historical record is full of periods of warming that take more than one step up, but if I had to guess I would say that we are currently experiencing the warmest period the Earth will have for a very, very long time.

Here is a close-up of the energy gap from the Eemian and the Holocene. The red line is the matched point of the rate of loss and 65N insolation. I don't predict the future from today, but I know which way I am betting.

The real question is, has the change in atmospheric CO_2 caused by the combustion of fuels been enough to counteract the behavior of the Earth/Sun orbital energy balance that has existed for the past million years. My very scientifically-based answer is "not a chance." There is no way for carbon dioxide to make up that energy gap. It is far too small a player.

It doesn't even matter that the 65N insolation is not going to drop as low over the next 10,000 years. The Earth cannot maintain the current warm temperature with even the current level of insolation. That is why the glaciers are growing ever closer to the equator. It takes time for a full glacial to develop and mankind is going to have a chance to watch it develop.
It is my hope that we as a species make the proper preparations.

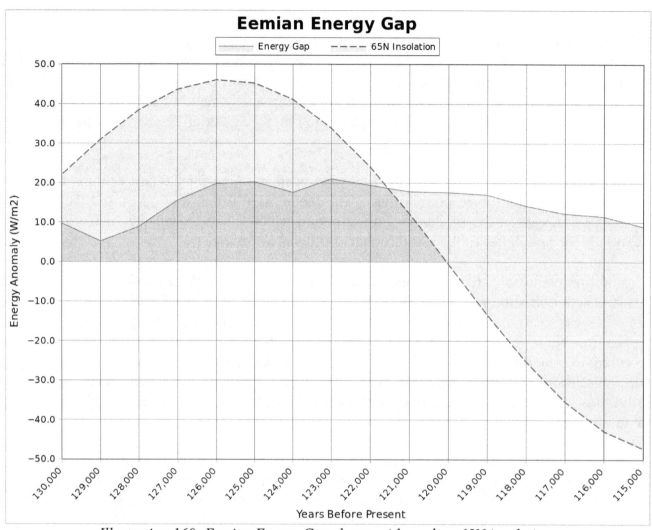

Illustration 160: Eemian Energy Gap shown with modern 65N insolation.

It is not just the dropping insolation that caused the Eemian to cool down, but the Earth as a whole and the NH in particular was losing too much energy when the insolation dropped. It is that combination of events that caused the Earth to cool down and trigger one of the longest glacials in the current ice age.

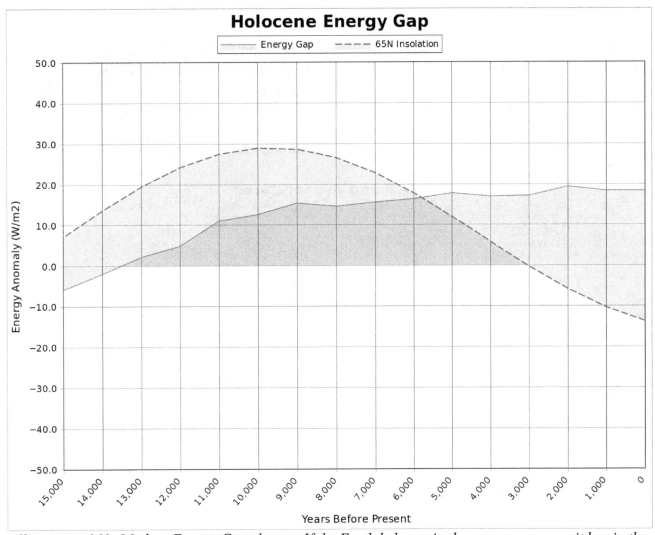

Illustration 161: Modern Energy Gap shown. If the Earth behaves in the same manner as it has in the past, the only possibility is long-term cooling.

The similarity between the current day and the Earth 119,000 years ago is amazing. In both cases the Earth was losing a similar about of energy to space while the 65N insolation was continuing to drop. The only significant difference is that the insolation was dropping faster during the Eemian, so the temperature should drop more slowly during the Holocene. The idea that the Earth will warm up against the natural drop that is taking place is very unlikely.

Observations of the Theory:

Illustration 162 shows the observed relationship between OLR and actual temperature. It is easy enough to also provide the theoretical slope of the line that shows the relationship between OLR and temperature. The measured OLR when the Earth is 13.5 °C is ~211.5 W/m2. That correlates to a transmission temperature of 233.6K. From that starting point it is a simple matter to determine the theoretical OLR for the Earth's temperature as it warms up. The result between the regression and the theoretical increase in OLR is nothing less than stunning.

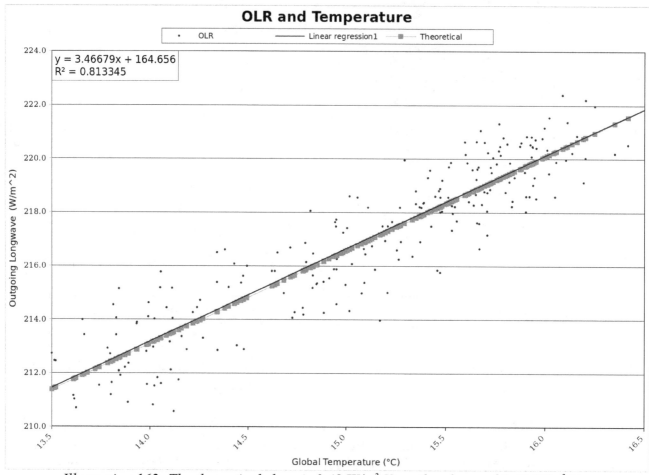

Illustration 162: The theoretical slope is 3.48 W/m² K vs. the observed 3.47 W/m² K.

The correlation between the individual data and the theoretical OLR is > 0.9. For the Earth to warm up by 3.0 °C, the peak temperatures would have to approach 20 °C. The theoretical OLR at that temperature is 235 W/m². How is a small increase in a radiative flux going to allow the Earth to lose an extra 15 W/m² to space for thousands of years without cooling down? That answer to that is very simple: it can't.

Chapter 14 The Climactic Future

I could write many more chapters for this book. I could cover the things like Climategate and the idea that there is a consensus that all scientists are convinced that global warming is real. There is really no reason for that. The science I have presented will either stand on its own merit or not. It will be attacked relentlessly by some. I accept that and am prepared (hopefully enough) for that. It will be ugly because the ideas presented here challenge the status quo.

There is a lot of money resting on the Theory of Global Warming. Many scientists are granted significant amounts of research dollars from governments to study global warming. That is why there are so many papers being released on it. Peer-reviewed papers follow the research dollars and those research dollars are all about global warming these days.

I do not think that there is some great conspiracy about global warming. What happened is that a few influential people became convinced that CO_2 would alter the climate and in the end they won the early debate and the result was that momentum was on their side, for a while at least. There was also the impeccable timing of the 1998 El Nino which provided a glimpse of climate variability. That was enough to convince more.

In the end, there was no alternative explanation to the carbon dioxide theory of global warming to explain the warming that was taking place. Scientists want theories to either support or oppose. The only theory that was really presented about why the Earth has been warming up for the past 100 years is the one based on CO_2. So there was almost a forfeit on the side of the skeptics. There was no coherent theory to oppose global warming, so global warming won the early rounds by default.

Then it became politically incorrect to oppose global warming and that is when things got ugly. If you don't believe that political correctness is part and parcel of the university academics life, then you have not spent enough time on a university campus.

Back in February, 2011 an article was published in the New York Times about a social science conference that took place in Texas. At that conference Dr. Jonathan Haidt asked the three following questions to the audience.

Question #1: How many consider themselves Democrats? Approximately 80% raised their hands. (800 of 1,000).

Question #2: How many consider themselves centrists or libertarian? Less than 36 people raised their hands. < 3.6%

Dr. Haidt the pointed out that such a distribution was "a statistically impossible lack of diversity." Such lack of diversity is especially common in the social sciences (just ask one, they will agree), but it is also true in academia in general although maybe not to the same degree.

Such a lack of diverse perspective makes group-think[163] all the more likely. More than anything I believe that the global warming debate became an issue of group-think. Fitting in became more important than improving the science.

Group-think is not one of my strong points; just ask anyone I work with. I have learned to be more tactful, but as a good engineer I am always opposed to what I view as bad ideas. When I started really looking into global warming I did not have a strong opinion other than I wanted to understand it for myself. So I started finding all the sources of data that I have used to put this book together.

I can say that I gave global warming a fair chance. It simply did not explain the behavior of the Earth's climate. While each and every swing of CO_2 can be explained by the temperature dependence on the solubility of CO_2 in water, there is not a good correlation to the global temperature and the level of CO_2 in the atmosphere. It simply failed that test in every past interglacial.

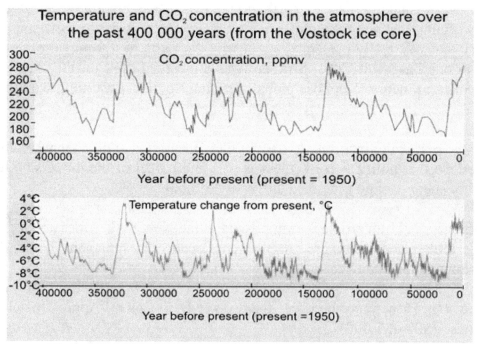

Illustration 163: It is charts like this that convince people that global warming is possible.

The only useful statement about the correlation of CO_2 and global temperature is that higher

CO_2 levels indicate the oceans are warm and lower levels indicate cooler oceans. That is all that the truly infamous correlation charts are capable of showing. The trick to these charts is they show them next to each other and they never ever overlay them. The very first time I overlaid the charts it was obvious to me that there was something missing in the explanation.

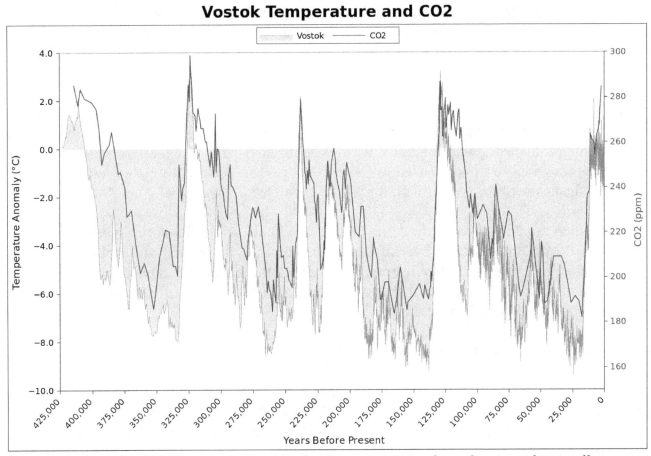

Illustration 164: The same data, but overlaid. That temperature drops first is evident in all cases.

In every case CO_2 rises after temperature. While warmists have been dealing with that issue for a long time, it simply doesn't matter. That the Earth has ALWAYS cooled while the CO_2 levels where high, says more about the Earth's climate than anything else. CO_2 has never once stopped the Earth from cooling down. If CO_2 caused the warming, it would prevent the cooling. That the Earth experienced thousands of years of cooling while CO_2 levels remained high was critical.

That is the single step down the path of becoming a skeptic. I had simply looked at the same chart that has been used to convince so many about global warming, but I looked at it in a different way. What I noticed was something that isn't normally discussed. The result was that my conclusions were different from looking at the same data.

It is possible that new evidence could prove my conclusions wrong. I accept this possibility as real, even if I think it is unlikely. What I do not accept, is that I am anti-science or anti-

265

environment. Reaching an unpopular conclusion has always been an important part of the scientific debate. Comparing skeptics to flat-earthers[164] is not part of a scientific debate, it is part of a smear campaign. Say whatever you like about my conclusions, but don't waste your time saying that I didn't look at the science.

Doing an independent review like this is what science is all about. That being a skeptic gets me labeled as anti-science is the greatest failing in the global warming debate. Anyone who reads this book is free to reach different conclusions from me and I will respect that. I might disagree, but that is part of a scientific debate. To denigrate the opposition is a tactic of political debate, not scientific debate.

There are skeptics who have a lack of scientific knowledge, but the solution that the warmists have taken is not to present the science in a straightforward manner, but to insult their lack of understanding and treat genuine inquiry with extreme disregard. I experienced this first hand many times long before I had reached my own conclusions. Asking a tough question is a good way to get the answer "it is too difficult for you to understand."

More than anything it was my desire to put the whole scientific story together as a single source that created this book. Even if someone reads this book and still believes in global warming, I will be happy if they learned more about the Earth and how it behaves. I have presented a version of the climate that is not dependent on the level of CO_2 in the atmosphere. This is something that is difficult to find.

In a way you could say that this is a carbon-free book. ;-)

It is time to spend one final moment with our settlers in Manhattan, Kansas. Let me move them a few years forward to October of 2010. That October in Manhattan was very warm. The average high temperature was 23 °C. That is 1.6 °C warmer than normal for the month. If the settlers had been there last year it might have seemed like a really good time to plant crops. It almost might make sense to believe that winter would not come. There were plenty of weeks that showed a warming trend. It was simply a warm month.

It also did not last. The Earth cooled down precisely like it should have, but instead of a mild winter, it was a brutal winter. Low temperatures reached below -20 °C and even some high temperatures were -10 °C. If they had not prepared for winter, they would have found themselves in a very bad position to survive.

Manhattan, KS 2010-2011

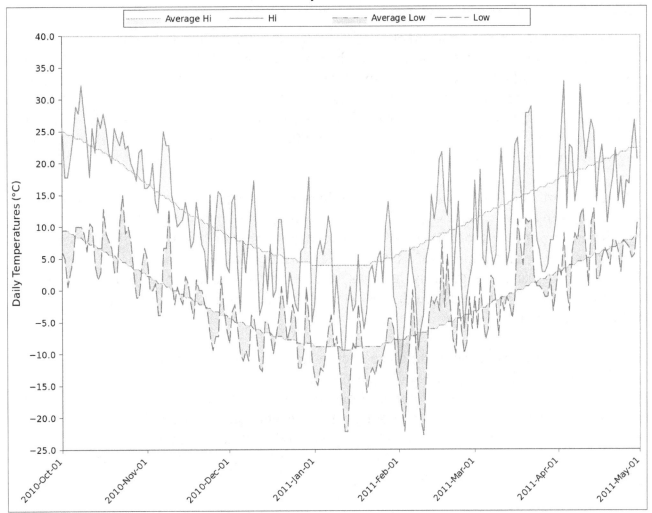

Illustration 165: The filled area is the daily temperature anomaly. (Red) High temperature, (Blue) Low Temperature

Everyone understands the yearly temperature cycle. It is predictable because we experience it many times. We know what to expect because we have seen it before. It becomes mundane the more we experience it. Eventually we only remember the exceptional events in the weather.

If our lives were on the order of millions of years, we would not be ignoring the indicators. We would have a very good idea what the climate would be like in a mere few thousand years. It is possible to know what the climate will be like in the future. It simply requires looking at the past in the correct manner.

There are only two options for the future of the Earth's climate. If CO_2 is the dominant factor, then the current CO_2 level combined with the upward temperature trend might be sufficient to break the Ice Age cycle that has been ongoing in the northern hemisphere for the last 2.6 million years and the Earth will warm up, the sea levels will rise maybe 10m and the ice age of the NH will end. This is by far preferable to the other option.

That option is also so unlikely that it is almost impossible to happen. CO_2 simply plays far too small a part in keeping the Earth warm. It's ability to cause small changes is really insignificant in the scale of the Earth's energy balance.

The natural and much more likely option is that the negative 65N summer insolation will slowly cause the developing glacial to strengthen over the next 1,000-3,000 years. Long before the glacial is actually fully developed, most of Canada, Northern United States, Europe and Russia will be experiencing extended periods of colder summers and occasional periods of summer frost. That will prevent most crops from growing in those regions. That the 65N summer insolation anomaly will be negative for the next 20,000 years is far more concerning to me than a change of 90 ppm of atmospheric CO_2.

That is the true nightmare scenario. What if half the crops in those regions were lost one year, then two years in a row? Then 5 years in a row? 10 years in a row? At what point will money no longer be enough to buy food. In 1816 it only took one summer to cause massive disruptions to the food supply. The population is much higher now and more people are dependent on purchasing food instead of growing it themselves. That is balanced by a much more efficient distribution system. In the end though, too many people depend on too much land that will not be able to produce food.

Studying the cycles of the Earth make it clear that the current climate will change. Nothing humanity does will alter that. The AGW period is already known to be smaller than the previous MWP. Already it is causing the Outgoing Longwave Energy to increase. As the OLR increases the Earth will experience even more cooling. The current warming period will come to an end. Maybe not in the next decade or two, but it will end. When it does humanity will struggle to feed itself more than ever before. Preparing for winter in the NH has always been an important part of life there. In that regard, nothing has changed...

Summer is Over...

Fall is Fading......

and Winter? Winter is coming!

Appendix 1: Glossary

65N – The 65N latitude. In this book it refers to the insolation that the 65N region gets during the summer months.

AGW – Anthropogenic Global Warming. The theory that the CO_2 plays a critical role in regulating the Earth's temperature. Since mankind is emitting CO_2, the theory is that this increased level of CO_2 will cause Global Warming.

AMO – Atlantic Multidecadal Oscillation. A natural oscillation of relatively warmer or cooler water in the North Atlantic. Cycle is ~70-90 years long.

Aphelion – The farthest distance of the Earth's orbit from the Sun. Currently this takes place in July.

Blackbody Temperature of the Earth – The theoretical temperature of the Earth if it had no atmosphere. The commonly accepted value is -19 °C (254K).

Climate Cycle – The current natural cycle of glacials and interglacials caused by the Milankovitch Cycle.

Convection – As used in this book. The energy transferred by natural convection from the Earth's surface to the atmosphere.

CO_2 – Carbon Dioxide. A naturally occurring gas in the Earth's atmosphere.

CRU – Climate Research Unit. Part of the University of East Anglia. CRU is used to describe the longest of the station methods of measuring the Earth's temperature anomaly. In their own terminology it is land air temperature anomalies on a 5° by 5° grid-box. Anomaly provided from 1850-present.

Eccentricity – The variation in how circular the Earth's orbit is over time. 400,000 year cycle.

ENSO – El Nino Southern Oscillation. A warm/cool oscillation of the equatorial water of the Pacific Ocean. Each phase tends to only last less than a year.

Forcing – also radiative forcing. See Radiative Flux. Forcing is used in place of radiative flux by many for the infra-red transmission from gases and clouds in the atmosphere.

GHE – Greenhouse Effect. The difference between the Earth's actual temperature and the Blackbody Temperature. The commonly accepted value is 33 °C.

GHCN – Global Historical Climatology Network. One of the station methods of measuring the Earth's temperature anomaly. It includes land and ocean measures of temperature. Anomaly provided from 1880-present.

Glacial – An interval of time within an ice age that is marked by colder temperatures and glacier advances.

Global Warming – As used in this book it refers specifically to Anthropogenic Global Warming (AGW).

Ice Age – A time of extensive glaciation covering vast areas of the Earth.

Insolation – The energy in W/m^2 that the Earth receives from the Sun.

Interglacial – Is an interval of time within an ice age that is marked by warmer temperatures and glacier retreats.

Latent Heat – As used in this book, the energy transferred by the evaporation of water at the Earth's surface to the atmosphere when the water vapor condenses into liquid or solid in the atmosphere.

LGM – Last Glacial Maximum. The coldest period of the last climate cycle that took place ~20,000 YBP.

LIA – Little Ice Age. A period between 200-400 YBP when the NH experienced a cooler climate.

Milankovitch Cycle – The theory that the varying orbital parameters (eccentricity, precession and obliquity) cause the Earth to experience the glacial/interglacial cycles by varying the energy at 65N during the summer months.

MWP – Medieval Warming Period. A controversial period that took place ~800-1,000 YBP when Greenland and other places in the North Atlantic were warmer than they are today.

NH – Northern Hemisphere of the Earth.

Obliquity – The 41,000 year cycle of the Earth's tilt. The tilt varies between 21-25 degrees.

PDO – Pacific Decadal Oscillation. A SST pattern in the North-Western Pacific Ocean of either relatively warmer or cooler temperatures. The west coast of the US tends to be the opposite phase.

Perihelion - The closest distance the Earth's orbit from the Sun. Currently this takes place in January.

Precession – The 26,000 year cycle of where the Earth is in it's orbit with respect to Aphelion and Perihelion.

ppm – Parts per million. As used in this book it is equal to ppmv.

Radiative Flux – The energy that is transmitted as infra-red energy by all objects, including gases.

RHT – Radiative Heat Transfer. The transfer of energy by means of radiative flux. The amount of energy transferred is dependent primarily on the temperature of the two objects. Warmer objects transfer energy to cooler objects.

RSS – Remote Sensing Systems. One of the two satellite measures of the Earth's temperature.

SH – Southern Hemisphere of the Earth.

Stable Isotope – Generally refers to the ratio of different isotopes of oxygen that are used to reconstruct historical temperatures. Hydrogen and carbon are two others that can be used.

Stratosphere – The middle portion of the atmosphere where the ozone layer resides. It is warmed by the oxygen/ozone cycle absorbing UV energy from the Sun.

Troposphere – The lower 12km of the atmosphere. This contains most of the total mass of the atmosphere and almost all of the water vapor in the atmosphere. It is warmed by energy being transferred from the Earth's surface.

TSI – Total Solar Insolation. The total energy that the Earth gets from the Sun. Only eccentricity affects the TSI.

OLR – Outgoing Longwave Radiation. Term used to describe the energy that the Earth is always losing to the surrounding space.

UAH – University of Alabama in Huntsville. One of the two satellite measures of the Earth's temperature.

W/m^2 – Units of energy per square meter. Compare to a 1x1 m solar panel. How much energy in those units would be the power provided by the solar panel.

YBP – Years before present. Specifically, years before the year 2,000 A.D.

Appendix 2: List of Charts, Graphics and sources used in the book

Illustration 1: The behavior of the Earth's temperature.
Illustration 2: Temperature and temperature anomaly shown
Illustration 3: The Earth's tilt causes the seasons. Picture from NOAA Online School.
Illustration 4: Winter Solstice. Daylightmap.com
Illustration 5: Summer Solstice. Daylightmap.com
Illustration 6: Northern Hemisphere is shown. SciencePhotoLibrary.com
Illustration 7: Southern Hemisphere is shown SciencePhotoLibrary.com.
Illustration 8: Arctic Circle: Mostly ocean surrounded by land. SciencePhotoLibrary.com
Illustration 9: Antarctic Circle: Mostly land surrounded by ocean Wikipedia Commons
Illustration 10: Tropics
Illustration 11: Different sources of the Earth's temperature data.
Illustration 12: The Blended temperature set.
Illustration 13: The Earth 50 Million Years Ago. Dr. Ron Blakey.
Illustration 14: The epochs of the .... Cenozoic. UC Berkeley.
Illustration 15: 30 Million years of the Earth's temperature. (Zachos, 2008)
Illustration 16: Opening of the Drake Passage Wikipedia Commons.
Illustration 17: Change in temperature when Antarctica transitioned to ice age. (Zachos, 2008)
Illustration 18: Antarctica has experienced wildly variable climate in the past. (Zachos, 2008)
Illustration 19: The last 65 million years. The last million are the coldest. (Zachos, 2008)
Illustration 20: Temperature history of the last 3 million years. (Raymo, 2005)
Illustration 21: Eemian temperature reconstruction from the EPICA ice core.
Illustration 22: Two different sources of the benthic stable oxygen isotope data
Illustration 23: PETM. (Zachos, 2008)
Illustration 24: Last eight climate cycles (Raymo, 2005)
Illustration 25: Weather.com monthly average for Manhattan, Kansas.
Illustration 26: EPICA ice core for the Eemian and Holocene interglacials. .
Illustration 27: Day-to-day temperature changes.
Illustration 28: Standard deviation of the monthly average temperature.
Illustration 29: Modern Day Tree Line. Source: Map The M Factory Smithsonian Institution
Illustration 30: The EPICA and Vostok ice core temperature reconstructions for the Eemian.
Illustration 31: Hippopotamus Wikipedia Commons
Illustration 32: Water Buffalo Wikipedia Commons
Illustration 33: Polar Bears Courtesy of the USGS
Illustration 34: Momma bear and cub. Courtesy of the USGS
Illustration 35: CO_2 levels dropped thousands of years after temperature
Illustration 36: Heavy Oxygen Cycle: NASA Earth Observatory
Illustration 37: CO_2 level as measured in the mid-western United States in 2006
Illustration 38: Mid-latitude insolation trend in the Autumn
Illustration 39: Normalized Energy from Sun
Illustration 40: Annual Energy and the temperature response.
Illustration 41: Hemispheric seasonal temperatures.
Illustration 42: Milutin Milankovic

1 http://nssdc.gsfc.nasa.gov/planetary/factsheet/earthfact.html
2 http://lwf.ncdc.noaa.gov/cmb-faq/anomalies.html#mean
3 Station method: http://www.cru.uea.ac.uk/cru/data/temperature/#faq
4 Satellite method: http://www.remss.com/msu/msu_data_description.html#rss_msu_data_analysis
5 CRU: http://www.cru.uea.ac.uk/cru/data/temperature/
6 GHCN: http://www.ncdc.noaa.gov/ghcnm/time-series/
7 UAH: http://vortex.nsstc.uah.edu/data/msu/t2lt/uahncdc.lt
8 RSS: http://www.remss.com/data/msu/monthly_time_series/
9 Krakatoa cooling: http://theinconvenientskeptic.com/wp-content/uploads/2011/03/Krakatoa-Mass_Portman_89.pdf
10 http://www.cefns.nau.edu/Academic/Geology/people/Dr.RonaldC.Blakey.shtml
11 http://cpgeosystems.com/products.html
12 Trends, Rhythms, and Aberrations in Global Climate 65 Ma to Present, James Zachos, *27 April 2001: Vol. 292 no. 5517 pp. 686-693*
13 http://www.usgs.gov/global_change/glaciers/glaciers_sea_level.asp
14 http://en.wikipedia.org/wiki/Geologic_time_scale
15 The Chicxulub Asteroid Impact and Mass Extinction at the Cretaceous-Paleogene Boundary. Schulte et al, Science 5 March 2010: Vol. 327 no. 5970 pp. 1214-1218
16.   Zachos, J., et al. 2008.
       Cenozoic Global Deep-Sea Stable Isotope Data.
       IGBP PAGES/World Data Center for Paleoclimatology
       Data Contribution Series # 2008-098.
       NOAA/NCDC Paleoclimatology Program, Boulder CO, USA.
17 Subtropical Arctic Ocean temperatures during the Palaeocene/Eocene thermal maximum, Sluijs. A. *Nature* **441**, 610-613 (1 June 2006)
18 http://www.esrl.noaa.gov/psd/data/gridded/data.ncep.reanalysis.html
19 Paleotemperature History of the Cenozoic and the Inititation of Antarctic Glaciation: Oxygen and Carbon Isotope Analyses in DSDP sites 277, 279 and 281, by N J Shackleton, JP Kennett
20 Eocene continental climates and latitudinal temperature gradients, Greenwood, David R. , Wing, Scott L.
21 Possible role of oceanic heat transport in Early Eocene climate, Sloan. L. Cirbus, PALEOCEANOGRAPHY, VOL. 10, NO. 2, PP. 347-356, 1995
22 Subtropical Arctic Ocean temperatures during the Palaeocene/Eocene thermal maximum, Sluijs. A. *Nature* **441**, 610-613 (1 June 2006)
23 The Antarctic Circumpolar Current System, Stephen R. Rintoul, Chris Hughes and Dirk Olbers
24 Zachos, J.C.; Kump, L.R. (2005). "Carbon cycle feedbacks and the initiation of Antarctic glaciation in the earliest Oligocene". *Global and Planetary Change* **47** (1): 51-66.
25 The Late Eocene-Oligocene Extinctions, Prothero, D. R., Annual Review Of Earth And Planetary Sciences, Volume 22, pp. 145-165
26 Gradstein, Felix; et al (2004). *A Geologic Time Scale 2004*. New York: Cambridge University Press. pp. 412
27   L. E., and M. E. Raymo (2005), A Pliocene-
      Pleistocene stack of 57 globally distributed benthic d18O records,
      Paleoceanography,20, PA1003, doi:10.1029/2004PA001071.
28 Exxon Sea Level Curve
29 http://earthobservatory.nasa.gov/Features/Paleoclimatology_OxygenBalance/
30 Mark Siddall. Palaeoclimate: The riddle of the sediments. *Nature* **437**, 39-41 (1 September 2005) | :10.1038/437039a; Published online 31 August 2005
31 Kennett, J. P.; Stott, L. D. (1991). "Abrupt deep-sea warming, palaeoceanographic changes and benthic extinctions at the end of the Palaeocene". *Nature* **353**: 225–229.
32 Sluijs, A.; Schouten, S.; Pagani, M.; Woltering, M.; Brinkhuis, H.; Damsté, J.S.S.; Dickens, G.R.; Huber, M.; Reichart, G.J.; Stein, R.; Others, (2006). "Subtropical Arctic Ocean temperatures during the Palaeocene/Eocene thermal maximum". *Nature* **441** (7093): 610–613
33 Katz, M.E.; Cramer, B.S.; Mountain, G.S.; Katz, S.; Miller, K.G. (2001). "Uncorking the bottle: What triggered the Paleocene/Eocene thermal maximum methane release" (PDF). *Paleoceanography* **16** (6): 667
34L.   E., and M. E. Raymo (2005), A Pliocene-
      Pleistocene stack of 57 globally distributed benthic d18O records,
      Paleoceanography,20, PA1003, doi:10.1029/2004PA001071.
35 http://grace.jpl.nasa.gov/data/pgr/

36 http://www.weather.com/weather/wxclimatology/monthly/graph/USKS0358

37   EPICA Dome C Ice Cores Deuterium Data.
        IGBP PAGES/World Data Center for Paleoclimatology
        Data Contribution Series # 2004-038.
        NOAA/NGDC Paleoclimatology Program, Boulder CO, USA.

38 http://www.wunderground.com/history/airport/KMHK/2006/10/18/MonthlyHistory.html#calendar

39 http://www.wunderground.com/history/airport/KBOI/2009/10/18/MonthlyHistory.html#calendar

40 http://www.nature.com/nature/journal/v431/n7005/abs/nature02805.html

41 http://www.athropolis.com/map5.htm

42 Eemian and Late Glacial/Holocene palaeoenvironmental records from permafrost
sequences at the Dmitry Laptev Strait , Sebastian Wetterich 2009

43 The Eemian stage in the Netherlands: history, character and new research, J.H.A. Bosch, P. Cleveringa & T.Meijer,
    Netherlands Journal of Geosciences 79 (2/3): 135-145 (2000)

44   EPICA Dome C Ice Cores Deuterium Data.
        IGBP PAGES/World Data Center for Paleoclimatology
        Data Contribution Series # 2004-038.
        NOAA/NGDC Paleoclimatology Program, Boulder CO, USA.

45   Vostok Ice Core Data for 420,000 Years, IGBP PAGES/World Data Center
        for Paleoclimatology Data Contribution Series #2001-076.
        NOAA/NGDC Paleoclimatology Program, Boulder CO, USA.

46 See Chapter 12.

47 http://www.hippoworlds.com/

48 http://www.pbs.org/wnet/nature/episodes/arctic-bears/how-grizzlies-evolved-into-polar-bears/777/

49 This section was initially published on The Inconvenient Skeptic Website.
    http://theinconvenientskeptic.com/2010/10/ice-core-data-truths-and-misconceptions/

50 http://earthobservatory.nasa.gov/Features/Paleoclimatology_OxygenBalance/

51 Park Falls CO_2, 2006: ftp://ftp.cmdl.noaa.gov/ccg/co2/GLOBALVIEW/gv_co2/lef010_01P2_ext.co2

52 ftp://ftp.ncdc.noaa.gov/pub/data/

53 http://earthobservatory.nasa.gov/Features/Milankovitch/

54 http://www.homepage.montana.edu/~geol445/hyperglac/time1/milankov.htm

55 Berger, A., 1992, Orbital Variations and Insolation Database.
    IGBP PAGES/World Data Center for Paleoclimatology
    Data Contribution Series # 92-007.
    NOAA/NGDC Paleoclimatology Program, Boulder CO, USA.

56 Picture from http://earthobservatory.nasa.gov/Features/Milankovitch/Images/orbital_variation.gif

57 Obliquity picture by Dennis Nilsson. Wikipedia Commons

58 Precession chart by Greg Benson. https://sites.google.com/site/bensonfamilyhomepage/Home/ice-age-and-global-
    warming   under the *GNU Free Documentation License*

59 EPICA Dome C Ice Cores Deuterium Data.
    IGBP PAGES/World Data Center for Paleoclimatology
    Data Contribution Series # 2004-038.
    NOAA/NGDC Paleoclimatology Program, Boulder CO, USA.

60 See Science Section for statistical behavior of the Eemian Autumn.

61 http://geology.com/press-release/gamburstev-mountains/

62 http://grace.jpl.nasa.gov/data/pgr/

63 Comparison between Holocene and Marine Isotope Stage-11 sea-level histories., Rohling, E.J., Braun, K., Grant, K.,
    Kucera, M., Roberts, A.P., Siddall, M., and Trommer, G, Earth and Planetary Science Letters, 291, 97-105

64 Ventilation of the Deep Southern Ocean and Deglacial CO2 Rise, Skinner L. C., Science 28 May 2010, Vol. 328 no.
    5982 pp. 1147-1151

65 Vostok Ice Core

66 The role of terrestrial plants in limiting atmospheric CO2 decline over the past 24 million years, Pagani. Mark, *Nature*,
    July 2, 2009.

67 http://www.esd.ornl.gov/projects/qen/nercAFRICA.html

68 National Geophysical Data Center

69 http://researchnews.osu.edu/archive/sajama.htm

70 Comparison between Holocene and Marine Isotope Stage-11 sea-level histories., Rohling, E.J., Braun, K., Grant, K., Kucera, M., Roberts, A.P., Siddall, M., and Trommer, G, Earth and Planetary Science Letters, 291, 97-105

71 Finlayson, Clive. Rapid ecological turnover and its impact on Neanderthal and other human populations. Trends in Ecology & Evolution. Volume 22, Issue 4, April 2007, Pages 213-222

72 http://www.guardian.co.uk/environment/2006/sep/26/conservationandendangeredspecies.climatechange

73 Comparison between Holocene and Marine Isotope Stage-11 sea-level histories., Rohling, E.J., Braun, K., Grant, K., Kucera, M., Roberts, A.P., Siddall, M., and Trommer, G, Earth and Planetary Science Letters, 291, 97-105

74 Anders E. Carlsona, David J. Ullmana, Faron S. Anslowc, Feng Heb, Peter U. Clarkd, Zhengyu Liub and Bette L. Otto-Bliesnere. Modeling the surface mass-balance response of the Laurentide Ice Sheet to Bølling warming and its contribution to Meltwater Pulse 1A. Earth and Planetary Science Letters July 2011.

75 See Science Section for Lost Cities.

76 Church. Estimates of the Regional Distribution of Sea Level Rise over the 1950–2000 Period. J. Climate, 17, 2609–2625.

77 http://sealevel.colorado.edu/

78 http://en.wikipedia.org/wiki/Flood_myth

79 REGIONAL SIGNATURES OF CHANGING LANDSCAPE AND CLIMATE OF NORTHERN CENTRAL SIBERIA IN THE HOLOCENE. Institute of Forest, Siberian Branch of the RAS, Akademgorodok, Krasnoyarsk, Russia Krasnoyarsk State Pedagogical University, 89 ul. A. Lebedevoi, Krasnoyarsk, Russian geology and geophysics, 2004, N 6, v. 45, p. 672-685

80 http://www.ncdc.noaa.gov/paleo/icecore/greenland/greenland.html

81 Frozen Annals Greenland Ice Sheet Research., Dansgaard W. Odder, Denmark: Narayana Press. pp. 124

82 http://www2.gi.alaska.edu/ScienceForum/ASF16/1698.html

83 http://www.ncdc.noaa.gov/paleo/icecore/trop/mtlogan/mtlogan.html

84 Moore, G.W.K., et al., 2002
Mount Logan Ice Core Isotope and Accumulation Data,
IGBP PAGES/World Data Center for Paleoclimatology
Data Contribution Series #2002-79.
NOAA/NCDC Paleoclimatology Program, Boulder CO, USA.

85 Schuster, P.F., et al., 2004,
Fremont Glacier 1991 Core Oxygen Isotope Data.
IGBP PAGES/World Data Center for Paleoclimatology
Data Contribution Series # 2004-079.
NOAA/NGDC Paleoclimatology Program, Boulder CO, USA.

86 http://www.ncdc.noaa.gov/paleo/icecore/trop/quelccaya/quelccaya_data.html

87 North Greenland Ice Core Project members. 2004.
North Greenland Ice Core Project Oxygen Isotope Data.
IGBP PAGES/World Data Center for Paleoclimatology
Data Contribution Series # 2004-059.
NOAA/NGDC Paleoclimatology Program, Boulder CO, USA.

88 http://en.wikipedia.org/wiki/Yonaguni_Monument

89 http://www.archaeologyonline.net/artifacts/gulf-of-cambay.html

90 http://climateaudit.org/2007/11/10/al-gore-and-dr-thompsons-thermometer-2/

91 http://pub.nsfc.gov.cn/pins_en/abstract/14/8/725.htm

92 http://www.britannica.com/EBchecked/topic/151663/Dark-Ages

93 Ingstad, Helge; Ingstad, Anne Stine (2001). The Viking Discovery of America: The Excavation of a Norse Settlement in L'Anse Aux Meadows, Newfoundland. Checkmark Books

94 Steig EJ, Hart CP, White JWC, Cunningham WL, Davis MD & Saltzman ES.
Changes in climate, ocean and ice sheet conditions in the Ross
Embayment at 6 ka. Annals of Glaciology 27, 305-310 (1998).

95 A quantitative high-resolution summer temperature reconstruction based on sedimentary pigments from Laguna Aculeo, central Chile, back to AD 850. von Gunten, L., Grosjean, M., Rein, B., Urrutia, R. and Appleby, P. 2009.; The Holocene 19: 873-881

96 Moberg, A., et al. 2005.
2,000-Year Northern Hemisphere Temperature Reconstruction.
IGBP PAGES/World Data Center for Paleoclimatology
Data Contribution Series # 2005-019.

NOAA/NGDC Paleoclimatology Program, Boulder CO, USA.

97 http://www.newscientist.com/data/images/ns/cms/dn11648/dn11648-2_726.jpg

98 http://www.farmersalmanac.com/weather/2010/03/22/the-year-without-a-summer/

99 Moberg, Anders . Highly variable Northern Hemisphere temperatures reconstructed from low- and high-resolution proxy data. Nature, Vol. 433, No. 7026, pp. 613 - 617, 10 February 2005.

100 http://vulcan.wr.usgs.gov/Volcanoes/Indonesia/description_tambora_1815_eruption.html

101 http://www.earlham.edu/~ethribe/web/tambora.htm

102 Oppenheimer, C. 2003. Climactic, environmental, and human consequences of the largest known historic eruption: Tambora volcano (Indonesia) 1815. Progress in Physical Geography 27:2. GeoREF. Accessed April 11, 2005

103 http://www.yourdiscovery.com/earth/year_without_summer/facts/index.shtml

104 Food vs. Fuel Debate. http://en.wikipedia.org/wiki/Food_vs._fuel

105 http://earthobservatory.nasa.gov/Features/Arrhenius/arrhenius.php

106 Cobb, K. El Niño/Southern Oscillation and tropical Pacific climate during the last millennium
Nature Vol. 424, No. 6946, pp. 271-276 (17 July 2003)

107 http://jisao.washington.edu/pdo/

108 http://www.aoml.noaa.gov/phod/amo_faq.php

109 Wanamaker, A.D., K.J. Kreutz, B.R. Schöne, N. Pettigrew, H.W. Borns, D.S. Introne, D. Belknap, K.A. Maasch, and S. Feindel. 2008. Coupled North Atlantic Slopewater forcing on Gulf of Maine temperatures over the past millennium. Climate Dynamics, 31(2-3), 183-194.

110 http://theinconvenientskeptic.com/2011/07/why-i-am-still-skeptical-about-sunspots-and-temperature/

111 http://mapcenter.hamweather.com/records/7day/us.html?c=maxtemp,mintemp,lowmax,highmin,snow

112 National Climatic Data Center, Asheville, N.C

113 http://www.usatoday.com/weather/wcstates.htm

114 http://www.ncdc.noaa.gov/oa/climate/globalextremes.html#hightemp

115 See Indian Summer in Science Content.

116 http://www.aspendailynews.com/section/home/147154

117 http://www.ncdc.noaa.gov/cmb-faq/anomalies.php

118 http://theinconvenientskeptic.com/2011/03/new-mystery-with-the-station-temperature-data/

119 http://science.nasa.gov/science-news/science-at-nasa/1997/essd06oct97_1/

120 See Chapter 12.

121 http://earthobservatory.nasa.gov/Features/EnergyBalance/page4.php

122 Earth's Global Energy Budget, Kevin E. Trenberth, John T. Fasullo, and Jeffrey Kiehl, Amercian Meteorological Society, 2008.

123 Science Section Chapter 12, Blackbody Temperature of the Earth.

124 See Chapter 12 for an extensive discussion of the atmosphere.

125 62% Cloudiness in Scientific Content.

126 See Chapter 2.

127 See Chapter 12 Scientific content.

128 Emigrant Pass Observatory - Insights on air and ground temperature tracking
Bartlett, M. G.; Putnam, S. N.; Chapman, D. S.; Harris, R. N.. American Geophysical Union, Fall Meeting 2003, abstract #PP52A-0959. 12/2003

129 Earth's Annual Global Mean Energy Budget
J. T. Kiehl and Kevin E. Trenberth
National Center for Atmospheric Research,* Boulder, Colorado

130 Dessler, A. E., Z. Zhang, and P. Yang (2008), Water-vapor climate feedback inferred from climate fluctuations, 2003–2008, Geophys. Res. Lett., 35, L20704, doi:10.1029/2008GL035333.

131 WATER VAPOR FEEDBACK AND GLOBAL WARMING[1]
Isaac M. Held and Brian J. Soden
Annual Review of Energy and the Environment, Vol. 25: 441 -475 (Volume publication date November 2000)

132 http://www.gly.uga.edu/railsback/FundamentalsIndex.html#Atmosphere

133 http://www.gly.uga.edu/railsback/FundamentalsIndex.html#Atmosphere

134 http://nssdc.gsfc.nasa.gov/planetary/factsheet/earthfact.html

135 http://www.gly.uga.edu/railsback/FundamentalsIndex.html#Atmosphere

136 See Energy Balance Chapter.

137 http://www.ozonelayer.noaa.gov/science/basics.htm

138 http://www.ccpo.odu.edu/~lizsmith/SEES/ozone/class/Chap_5/index.htm

139 http://gcmd.gsfc.nasa.gov/Resources/FAQs/ozone.html

140 If there was no oxygen/ozone in the atmosphere, the Earth's surface would absorb the energy instead of the stratosphere. This of course would significantly warm up the Earth's surface. So the oxygen/ozone in the stratosphere have an enormous cooling effect on the Earth's surface. So in this case the greenhouse gases have a shading effect on the Earth.

141 http://epw.senate.gov/public/index.cfm?FuseAction=Files.View&FileStore_id=84462e2d-6bff-4983-a574-31f5ae8e8a42

142 See Blackbody temperature of Earth in Science Section.

143 NASA, Blackbody of Earth.

144 http://www.asi.org/adb/m/03/05/average-temperatures.html This is a frustratingly difficult topic to pin down.

145 See Park Falls, Wisconsin in the same chapter.

146 http://en.wikipedia.org/wiki/Greenhouse_effect

147 http://www.esrl.noaa.gov/psd/cgi-bin/DataAccess.pl?DB_dataset=NCEP/DOE+AMIP-II+Reanalysis+%28Reanalysis-2%29+Daily+Averages&DB_variable=Upward+Longwave+Radiation+Flux&DB_statistic=Mean&DB_tid=30286&DB_did=59&DB_vid=1486

148 CO2 Doubling Section of Science Content.

149 http://nssdc.gsfc.nasa.gov/planetary/factsheet/earthfact.html

150 Radiative Non-Equilibrium at the Lunar Surface
     Brendan Hermalyn
     2006 Summer Institute on Atmospheric, Biospheric, Hydrospheric, and Solar and Space Plasma Sciences
     NASA/Goddard Space Flight Center, Greenbelt, M

151 See Chapter 5.

152 http://en.wikipedia.org/wiki/File:Olr_1979_1995_mean.png,
     http://www.esrl.noaa.gov/psd/data/gridded/data.interp_OLR.html

153 http://www.spectralcalc.com/info/about.php

154 Alley, R.B.. 2004.
     GISP2 Ice Core Temperature and Accumulation Data.
     IGBP PAGES/World Data Center for Paleoclimatology
     Data Contribution Series #2004-013.
     NOAA/NGDC Paleoclimatology Program, Boulder CO, USA.

155 See Science Content for details.

156 Lisiecki, L. E., and M. E. Raymo (2005), A Pliocene-Pleistocene stack of 57 globally distributed benthic d18O records, Paleoceanography,20, PA1003, doi:10.1029/2004PA001071.

157 http://www.esrl.noaa.gov/psd/data/gridded/data.interp_OLR.html

158 http://www.ipcc-data.org/ddc_co2.html

159 Variability of El Nino/Southern Oscillation activity at millennial timescales during the Holocene epoch. Christopher M. Moy, Geoffrey O. Seltzer, Donald T. Rodbell & David M. Anderson. NATURE | VOL 420 | 14 NOVEMBER 2002

160 A Neoproterozoic Snowball Earth. Paul F. Hoffman, Alan J. Kaufman, Galen P. Halverson and Daniel P. Schrag. Science 28 August 1998: Vol. 281 no. 5381 pp. 1342-1346

161 Science Content in Chapter 12.

162 http://www.esrl.noaa.gov/psd/data/gridded/data.interp_OLR.html

163 http://en.wikipedia.org/wiki/Groupthink

164 http://www.cbsnews.com/stories/2008/03/27/60minutes/main3974389.shtml

CPSIA information can be obtained
at www.ICGtesting.com
Printed in the USA
BVHW080112230419

546159BV00022B/1475/P